Essays in the Study of Scientific Discourse

Methods, Practice, and Pedagogy

ATTW Contemporary Studies in Technical Communication

M. Jimmie Killingsworth, Series Editor

Published in cooperation with the Association of Teachers of Technical Writing

Volume 1:
Foundations for Teaching Technical Communication: Theory, Practice, and Program Design
edited by Katherine Staples and Cezar M. Ornatowski, 1997

Volume 2:
Writing at Good Hope: A Study of Negotiated Composition in a Community of Nurses
by Jennie Dautermann, 1997

Volume 3:
Computers and Technical Communication: Pedagogical and Programmatic Perspectives
edited by Stuart A. Selber, 1997

Volume 4:
The Practice of Technical and Scientific Communication: Writing in Professional Contexts
edited by Jean A. Lutz and C. Gilbert Storms, 1998

Volume 5:
The Dynamics of Writing Review: Opportunities for Growth and Change in the Workplace
by Susan M. Katz, 1998

Volume 6:
Essays in the Study of Scientific Discourse: Methods, Practice, and Pedagogy
edited by John T. Battalio, 1998

Essays in the Study of Scientific Discourse

Methods, Practice, and Pedagogy

edited by
John T. Battalio
Boise State University

Volume 6 in ATTW Contemporary Studies in Technical Communication

Ablex Publishing Corporation
Stamford, Connecticut
London, England

Printed in the United States of America

Library of Congress Cataloging-in-Publication Data

Essays in the study of scientific discourse : methods, practice, and Pedagogy / edited by John T. Battalio.
p. cm. — (ATTW contemporary studies in technical communication ; v. 6)
Includes bibliographical references and index.
ISBN 1-56750-384-5 (cloth). — ISBN 1-56750-385-3 (pbk.)
1. Science—Methodology. 2. Science—Philosophy. I. Battalio, John T. II. Series
Q175.E736 1998
502.8—dc21 98–5848
CIP

Ablex Publishing Corporation
100 Prospect Street
P.O. Box 811
Stamford, CT 06904

Published in the U.K. and Europe by:
JAI Press Ltd.
38 Tavistock Street
Covent Garden
London WC2E 7PB
England

Contents

Introduction

John T. Battalio
Boise State University

Contemporary scholars of scientific discourse have benefited in no small part from the groundwork laid by scholarship and critical thinking in a number of fields.[1] Sociologists of science, including Thomas S. Kuhn (1970), Bruno Latour and Steve Woolgar (1979), Michael Mulkay (1979), and most notably Robert K. Merton (see, for instance, 1970, 1973; Merton & Gaston 1977), have shown the influence of the scientific community on the development of scientific knowledge.[2] Scholarship in the rhetoric of science, starting with Philip Wander's 1976 definition, has recently been surveyed by Dilip Parameshwar Gaonkar (1993).[3] Historians of science have placed scientific discourse in chronological perspective.[4] Philosophers of science have uncovered the epistemological complexity of scientific knowledge and its discourse,[5] and a number of scientists have added their own insights from within science as well.[6]

Since the mid 1980s, there have been a number of notable studies of scientific discourse, which add further insights into scientific thought, discourse, and activity. Bruno Latour (1987) follows scientists and engineers through the process of turning facts, methodologies, theories, and instruments into implicitly accepted entities. In doing so, he describes a complex, nearly impenetrable network of claims, counterclaims, associations, texts, citations, and mechanisms through which dissenters must pass before being heard. Charles Bazerman (1988), whose empirical methodology has been a model for numerous rhetoricians and historians of science, uses case studies of both scientific organizations and individuals to describe the evolution of scientific communities. Through rhetorical and linguistic analyses of the experimental research article, personal narratives, and social interactions, his expansive study supports a social constructionist model of science.

Proposing a link between rhetorical invention and the creation and progress of scientific discourse, Lawrence Prelli (1989) uses scientific *stasis* procedures for determining both how scientists can evaluate adverse community criticism and how they can formulate effective responses to that criticism. He also proposes a set of scientific *topoi*, which guide scientists' lines of argument. Alan Gross (1990) broadens classical rhetorical constructs in scientific discourse to include Aristotelian persuasion, tropes and analogies, and classical arrangement. In describing linguistic change, the interrelation between rational and rhetorical reconstructions, the importance of authority and power relationships in science, and the link between science and society, Gross advocates the relativity of science and the predominance of rhetoric in science, which ultimately uses persuasion to advance scientific knowledge.

Viewing texts as discourse to be interpreted, Greg Myers (1990) makes a number of observations about the construction and interpretation of scientific texts. He describes the building of consensus in proposal writing and the changing status of scientific claims during the publication process. To Myers, not only scientific narratives, but also popularizations of science create the cultural authority of scientists.

Rhetoricians have also expanded their studies to the processes involved in the transfer of scientific discourse to the public forum. In numerous articles analyzing popular science writing, scholars describe how the mass media has perpetuated the general public's idealistic metanarrative of science as an objective forum whose discoveries improve the world.[7] Others have begun to study the rhetoric of public discourse on the environment. In particular, Jimmie Killingsworth and Jacqueline Palmer (1992) describe the methods science uses to insulate itself from this discourse, leaving solutions for environmental dilemmas to engineers and public policy makers. The authors challenge science to evaluate its discourse according to a higher ethical standard whereby ecologists will more readily influence decision-making, acknowledge the relativity of scientific knowledge, and suggest modes of action during uncertainty.

M. A. K. Halliday and J. R. Martin (1993) use functional linguistics to highlight the lexico-grammatical features of scientific writing. The authors' premise is that the evolution of scientific grammar made possible the evolution of both scientific thought and, by extension, science itself. The means for this grammatical evolution was, first, the ability of language to nominalize and, second, the birth of grammatical metaphor, whereby multinomials enable one phenomenon to represent another. Through such objectification, science creates technical taxonomies, which enable scientists to interrelate objects and embed processes.

Marcello Pera (1994) synthesizes Aristotle, Chaim Perelman, and Thomas Kuhn, among others, to propose a dialectical model for scientific argumentation. As a central argumentative component, rhetoric assists scientists in choosing, interpreting, and applying a specific methodological rule; justifying initial premises; and supporting or discrediting scientific hypotheses. Pera identifies a number of rhetorical strategies used by scientists to determine the viability of rival theories.

Charles Alan Taylor (1996) traces the construction of contextual boundaries which separate science from other practices. Questioning the focus of previous studies of science on the nature of science and scientific facts and on the various spheres of argument from personal to public to technical, Taylor suggests as a more fruitful endeavor a study of the discursive, and thus rhetorical, practices contributing to the demarcation of science from other areas of endeavor. He models this methodology in two case studies of demarcation processes, one of nonscience (creationism) and the other of bad science (cold fusion).

Interest in the discourse of science has also prompted a number of journals to publish special issues on the rhetoric of science. In the *Southern Communication Journal* (Summer, 1993), Gaonkar critiques the writings of several contemporary rhetoricians of science. Questioning the universality of rhetoric, Gaonkar argues that rhetorical terminology, which was intended for analysis of oratory, is inadequate for studying other forms of discourse. Rhetoricians' responses to Gaonkar range from those who believe that rhetoricians of science should be more reflexive of their own theoretical approaches and methodology (Gross, 1993) to those who believe that Gaonkar's definition of rhetoric is too narrow (Prelli, 1993).

In addition to the two articles reprinted here, *Technical Communication Quarterly* (winter 1994) describes the rhetorical skill in one physicist's negotiation of two distinct scientific audiences; the persuasive, though biasing, effect of using large-scale trends in evolutionary biology; and the benefits to rhetoricians of studying risk-communication theory as exemplified in the debate over nuclear energy.

Three of the four articles in the *Rhetoric Society Quarterly*'s fall 1996 issue describe the limitations of scientific discourse. In its evolution from manufacturing quality control to quality management assurance, the field of quality control marginalized the discourse of pure science in favor of a discourse of business and the marketplace, thus reversing the traditional route to professionalization in which practitioner gives way to academician. A study of nuclear fusion research at one laboratory emphasizes the "laboratory as rhetor," through which scientists rearticulate their own goals in light of their sponsors'—a ploy which eventually proves unsuccessful because of a changing political scenario. A third article argues that limitations inherent in language and scientific methodology prevented the author from convincing brain-sex researchers to consider a gender-diversity hypothesis. Finally, rhetoricians' attention is turned from metaphor to the classical-rhetorical figures of *incrementum* and *gradatio* to describe series reasoning in scientific argument.

Studies of science writing have also appeared in several volumes of collected essays. With the goal of improving the conduct, content, and communication of rhetoric in scholarship and public affairs, the 22 essays in *The Rhetoric of the Human Sciences* (Nelson, Megill, & McCloskey, 1987) touch upon a wide range of disciplines from anthropology, psychology, economics, and political science to theology, law, and feminist studies. The volume has been influential in showing the wide range of subject matter available for rhetorical inquiry into scientific discourse.

Persuading Science (Pera & Shea, 1991) includes a number of theoretical arguments supporting the influence of persuasion in scientific argumentation as seen from the perspectives of the language of science, scientific dialectics, and theory acceptance. Other essayists give credence to these viewpoints by offering historical evidence from the works of a wide range of scientists, including Galileo, Descartes, Newton, Hobbes, and Einstein.

Social constructionist views permeate *Textual Dynamics of the Professions* (Bazerman & Paradis, 1991). Rhetorical studies of the writings of Joseph Priestley, and of Herbert Spenser, Thomas Huxley, and John Dewey, in addition to a study of review articles by two molecular biologists, show the importance of texts in the establishment of professions. Another section describing the enrollment of novices into professional communities has implications for scientific communities as well.

Two recently edited volumes consider the effects of culture on science. In *Science as Practice and Culture* (Pickering, 1992), contributors struggle with the relationship between scientific practice and culture on the one hand and the sociology of science on the other. Of particular interest to researchers of medical-science texts are two articles in *The Literature of Science* (McRae, 1993) analyzing the writings of practitioners in the fields of surgery and neurology, respectively, who turned to professional writing in order to offset the detachment inherent in the scientific tradition. In another article, Jeanne Fahnestock uses classical rhetoric to analyze the shift from the purely scientific article to scientific popularizations.

Finally, the fourteen essays in *Understanding Scientific Prose* (Selzer, 1993) are unique in that they all critique the same scientific article. Though scientific rhetoric and argumentation are analyzed, emphasis is on the variety of methods of analyses available to rhetoricians of science. Divided equally between contextual and textual analyses, the volume includes studies based on intertextuality; feminist, reader-response, and deconstruction theories; historicism; sociolinguistics; and classical rhetoric.

The essays in the present volume continue the perceptive analysis of scientific discourse modeled by their predecessors. Included here are empirical methodologies; literary-critical analyses; case studies; and linguistic, historical, and pedagogical perspectives. As shown by the brief overviews below, these essays are as rich and diverse as the rhetorical studies that precede and inform them.

RHETORICAL INTERPRETATIONS OF SCIENCE

In "Uncertain Science and the Sponsored-Research Process," Gay and Ted Gragson's case study shows how technological advances may increase the rhetorical complexity of the grant proposal process. The authors' study describes the awarding of a government grant for phytoremediation research—a process complicated by the uncertain effects of genetically engineering plants to process and discharge mercury from polluted soil. By restricting their study to government funding, the authors are able to pinpoint more clearly both the external and internal forces involved in decisions for the awarding of public funds. In the process, Gragson and Gragson show that contemporary "basic" science is not easy to define. They reveal the tension between two competing models of science: first, the traditional methodological model based on Newtonian empiricism and logical positivism (the Kuhnian puzzle-solving model); and, second, a newly emerging dialectical, semantic model, where decisions are based, not on what *is*, but on what *might be*. From the grant writers' perspective, the 10-year process toward full funding of their phytoremediation research taught them to de-emphasize the uncertainty inherent in their research, while emphasizing its place in an ongoing research process.

Christine Skolnik in "Neuropsychology, Sex, and Rhetoric" contrasts a scientific with a critical-rhetorical view of sexual difference. Through her descriptions of the neuropsychological research of Doreen Kimura and the theoretical approaches of Judith Butler, Skolnik shows that both studies, though methodologically distinct, advance ambiguous, fluid interpretations of human sexual identity. The chapter poses other questions for rhetorical studies of human sciences, as Skolnik points out the interpretive dilemma social scientists face. In drawing conclusions about societal norms, scientists must ultimately base their conclusions on studies of individual difference. As Skolnik comments, "Psychologists are aware of the causal chasm ... individuals are not assumed to inhabit norms." But, as her chapter shows, the stereotypical norms implicit in Kimura's studies are the very "conclusions" the popular press draws from her work. Ironically, however, in this ambiguity of methodology, Kimura's research describes the same human sexual ambiguity as does Butler's, who in contrast is informed by theorists like Foucault. One might wonder in conclusion how really far apart are rhetoric and science.

Mary Rosner's "Plotting a Middle Ground: An Account of a Failed Science Story" extends Skolnik's critique of gender roles to another level by juxtaposing feminine versus masculine assumptions about scientific objectivity and rationalism. Her case study analyzes the controversy surrounding scientist Evelyn Fox Keller's biography of geneticist Barbara McClintock. Keller, in defending her placement of McClintock within traditional, patriarchal science, proposes a view of science composed of a "stable middle ground" of domination, mastery, and control, where neither patriarchal nor feminist science distorts. Keller's failure to persuade her feminist critics, Rosner suggests, was due to Keller's own acceptance of only one interpretation of McClintock, a reflection of masculine science's objectivity, mastery, and certitude. Had Keller used Susan Leigh Star's "high tension zone," where "blurred boundaries, uncertainties, and instabilities" are valued for providing multiple perspectives, Rosner believes that Keller's feminist critics might have been persuaded. Rosner's study forces readers to reconsider their own understanding of the mechanisms involved in establishing scientific norms and community. To what extent does science acknowledge and act upon uncertainty, instability, and contradiction? If feminine and masculine science were to share equal authority, what would this restructured scientific world look like? Rosner further reflects upon the slipperiness of language itself. Given the concept of reader as meaning-maker, to what extent can scientific discourse reflect the laboratory? If language's multiplicity is central to its effectiveness, then how can scientific language be clear, precise, and unambiguous?

"Foucault's Archaeological Method and the Discourse of Science: Plotting Enunciative Fields" by Cynthia Haller shows how scientific claims change as they move from the scientific to the public arena. Employing Michel Foucault's archaeological methodology, Haller studies two discourse events on the subject of Chronic Fatigue Syndrome (CFS): a) the presentation of a CFS researcher before a group of his colleagues; and b) a *Newsweek* article describing the researcher's work. In forming objects, contrasting grids of specification emerge: those of scientific research versus those of personal experience. In forming subject positions, speakers play distinctive roles—one as a member of a well established discourse community, the other as a conqueror of good over evil. In organizing concepts, the style varies from quantitative to humanistic. In locating themes or strategies, Haller shows that apparently opposing discourse strategies are really complementary nodes in the power matrix emerging from the discourse events themselves. Still, the forces of power and knowledge also transcend the boundaries of these two very distinct rhetorical situations, as for instance, in the increased prestige for CFS researchers' work and an enhanced awareness of the status of CFS as a disease. Thus, Haller questions the dichotomy, which rhetoricians of science often make, between scientific discourse and popular writing about science.

A CASE STUDY OF A CONTEMPORARY CONTROVERSY IN SCIENCE

Contemporary science has provided rhetoricians of science a perfect model for study: The cold fusion controversy has revealed the many facets of scientific argumentation and persuasion, consensus building, and fact-finding. It is appropriate then that two recently published companion pieces on the controversy be reprinted here.

In "Cold Fusion and the Sociology of Scientific Knowledge," Trevor Pinch provides a lucid explanation of the scientific controversy itself, as well as a concise summary of the historical events from the announcement by Pons and Fleischmann that they had produced cold fusion to their experiment's subsequent discredit. In showing how both sides in the controversy defend their positions, Pinch describes the pressures of publication and the establishment of priority, the influence of replication, the effects of bypassing the peer-review process and the subsequent rise of an impromptu electronic review process, interdisciplinary rivalry, and the power relationships within a scientific discourse community. Clearly shown, then, is the rhetorical richness surrounding the controversy, as well as a case in point of science in action.

Charles Alan Taylor takes the controversy a step further by looking at "Science as Cultural Practice: A Rhetorical Perspective." He interprets the many complex issues of the cold fusion controversy as a manifestation of the cultural practices surrounding science. In addition to viewing science from its own knowledge claims and from the means whereby these claims are legitimized, scientific rhetoric can also be informed by analyzing events outside both the laboratory and scientific texts themselves. Such occurrences as press conferences, Congressional budget hearings, and grant application deliberations are part of the cultural practices which inform science and influence its final outcomes. Thus, because these external discourses are worthy of analysis, they appropriately belong under the umbrella of scientific rhetoric along with scientific texts themselves. Taylor points out that, in the case of the cold fusion controversy, internal scientific practices cannot entirely account for the end of the debate—external influences contributed as well.

HISTORY OF SCIENTIFIC DISCOURSE

In "'Practicing our Preaching': Gifford Pinchot, Conservation, and the Scientific Rhetoric of 'Demonstration Forestry,'" John Carpenter reveals the rhetorical power of the scientific report. Gifford Pinchot, head of the fledgling U. S. Bureau of Forestry, successfully used this genre to convince late 19th-century landowners and legislators of the technical and economic advantages of forest management. Expanding upon historian Samuel P. Hays' perception that the conservation movement was rooted in science rather than in the populace, Carpenter rhetorically analyzes two 1890s reports by Pinchot, who argues on behalf of a scientific approach to forestry. Carpenter demonstrates the complexity of the task: Pinchot had both to distance himself from the transcendentalist view of conservation, while at the same time to incorporate scientific practice into the methodology espoused by timber advocates. Early on, Pinchot's objective writing style in conjunction with his precise methodology argued in favor of forestry as a technical activity. Yet Carpenter points out Pinchot's careful filtering of information so as to give the forester an implicitly privileged role in the process. At the same time, however, Pinchot must mesh the goals of lumbering with those of professional forestry. Ultimately, the merger of rhetoric and the scientific report proved successful, as shown by the millions of acres of private forest land under government management during Pinchot's headship.

In "Rationality Reborn: Historical Roots of the Passive Voice in Scientific Discourse," Dan Ding's historical search carries him back to the Age of Reason and its influence on the transmission of scientific knowledge. He first briefly reviews the literature demonstrating the drastic shift toward the impersonal passive by the end of the 19th century. Then, reflecting on the writings of Bacon, Locke, Darwin, and Spencer, he attributes the change from active to passive to the influence of rationality and reason, which ultimately move science from competitive values and personal argumentation to public cooperation and communal identity. Ding gathers a number of components of scientific rhetoric under the umbrella of rationality and reason: the rise of scientific journals; changes in the use of citations; appearance of the collective "we"; refinement of scientific methodology; peer review; the route to scientific authority; implied and real audiences; and, most importantly for Ding's study, the passive voice. Implied in all of this, of course, is the notion of professionalization, which, significantly, Ding references in his closing remarks. That science may be said to have professionalized by the turn of the 20th century contributes in no small part to changes in scientific style.

LANGUAGE OF SCIENCE

Dwight Atkinson in "Integrating Multiple Analyses in Historical Studies of Scientific Discourse: *The Philosophical Transactions of the Royal Society of London*, 1675–1975" refines Bazerman's methodology by incorporating into rhetorical analysis a sociolinguistic methodology, multidimensional register analysis, as another means of contextualizing scientific discourse. The study employs over 200 texts published over a period of 300 years by the Royal Society's *Transactions*. The rhetorical analysis demonstrates the shift from author-centered to object-centered discourse, active to passive constructions, letter-based to report-based communication, and the rise of an identifiable discourse community. Register analysis reveals shifts from involved to informational production, narrative to non-narrative, and concrete to abstract information. Atkinson's analysis clarifies the interrelationship between these rhetorical and linguistic shifts, thus merging rhetorical and empirical methodology even further.

Richard Johnson-Sheehan surveys scholarship that studies the extent to which metaphor has influenced scientific theories and beliefs. Beginning with scholarly studies linking metaphor with thought, "Metaphor in the Rhetoric of Scientific Discourse" first traces two lines of research: studies identifying an expansive role for metaphor in structuring scientists' thought patterns; and those restricting metaphorical influence to changes in the literal lexicon used by scientists. In discussing the exact function of metaphor in scientific discourse, scholars have generally followed Kenneth Burke's insight that metaphor serves to guide or change perspective. However, scholars disagree as to how scientific metaphors actually work. Interaction theories, which contend that metaphor assigns new meaning or insight, have been most prominent, though Johnson-Sheehan points out the limitations of these approaches in his overview of more recent theories, which are based on audience invention and interpretation of metaphorical language. Finally, he observes that a number of rhetoricians have taken a utilitarian approach to scientific metaphor by focusing on how science uses metaphor. Here, scholars emphasize the power of metaphor to invent not just lexical meanings, but even entire theories. Through this power, scientists have the ability to reconceptualize scientific knowledge and empirical events from different perspectives.

In "The Language of Classifying in Introductory Science Texts," Steven Darian explores the multifaceted use of classification schema in scientific thought by analyzing the role of classification in two university-level texts. His study is based on a collection of sample sentences and phrases which exemplify the major forms of classification used by science. The chapter itself is a model of classification, as it categorizes the various syntactic patterns of classification, the lexical features used by his sample authors, and graphical representations which clarify, reinforce, and sometimes initiate classification schemas. His analysis shows the fluidity, flexibility, and slipperiness of these schemas, whereby some categories overlap at times and others merge. Implicit then in his analysis is that scientific knowledge itself may be as slippery as the classification schemas from which science often derives its own authority. Theoreticians of science, including Kuhn and Lyotard, have stressed the importance of university education in initiating newcomers into the scientific community. Darian's study clarifies the role of at least one element in this initiation process. His chapter is also a fitting transition to the final section on pedagogy, because he ends by suggesting several classroom activities for introducing students to the concept of classification.

PEDAGOGICAL APPROACHES TO THE DISCOURSE OF SCIENCE

By showing that changing editorial prescriptions and scientific audiences should influence the language of scientists, Philippa Jane Benson advocates a greater cooperation between scientists and instructors of writing. In "Changing Moorings in Scientific Writing: Suggestions to Authors, Allusions for Teachers," she compares both the changes from the fifth to the recently published sixth edition of the CBE style manual and the "Instructions to Authors" for the biological journals *Nature*, *Ecology*, and *Cell*. Changes in the CBE manual reflect the increasingly interdisciplinary character of the sciences, as well as their broadening international scope. These changes are also apparent in the editorial specifications of the three journals studied. As a result, scientific writers must provide meaningful contexts for both nonspecialists and nonnative speakers. Implicitly then, scientific writers need to accommodate these varied audiences by greater attention to such stylistic features as vocabulary, sentence structure, and sentence complexity. That is, audience analysis skills are now even more important for scientific writers, whose success in publication depends largely on how well they describe the importance of their research to their readers.

In "Textual and Contextual Models of Science Writing," Ramón Plo Alastrué and Carmen Pérez-Llantada describe the diversity of textual materials available for science-writing courses. Basing their choice of texts on sociolinguistic and social constructionist theories, the authors argue that content-based materials, such as the scientific article, abstract, thesis, and monograph, should be supplemented with pragmatic-based materials, such as citation indices, grant proposals, calls for papers, professional correspondence, and Internet discussion lists. The former introduce students to the specialized language of science, to the complexity of scientific content, and to accepted discourse formats through their role as textual models. The latter show students "pretexts," or "hypotexts"; that is, the discourse forerunners of scientific literature. Because content materials portray objective science, and pragmatic materials show science in process, students understand scientific discourse as part of a living, evolving discourse community. Thus, such a diverse collection of materials will teach students not only the scientific genres, but also the parameters which create the discourse—that is, the contexts of production *and* of use.

Finally, in "Practice-Centered Approaches to Literacy: Applications for the Research and Teaching of Science Writing," Rosemary Horowitz, reflecting on changing definitions of literacy, incorporates Dell Hymes' "speech event," Shirley Brice Heath's "literacy event," and Brian Street's "literacy practice" into her own practice-centered approach to the research and teaching of science writing. Focusing on literacy events and practices, her discussion integrates more traditional approaches to the study of science writing, such as audience analysis, authorship, graphical representation, scientific style, historical origins, and genre, into a larger sociolinguistic context. Horowitz's examination of four science texts shows the multiplicity of research and pedagogical questions arising from her practice-centered model. Her analysis reinforces the fact that the simple question of whether or not science writing can be interpreted in light of social practice has given way to a multiplicity of sociolinguisic and literacy questions—for instance, how to measure the social distance between writer and reader and among co-authors, what shared language norms help to create a scientific text, what set of literacy events and practice mold a particular scientific genre, and what communicative competency science writers need.

ACKNOWLEDGMENTS

To all of these contributors, I express my thanks not only for their contributions to this collection, but also for their patience and promptness in answering my many editorial queries. The following publishers and journals kindly granted permission to reprint excerpts or articles from their publications:

Atkinson, Dwight S. (1996). The *philosophical transactions of the Royal Society of London*, 1675–1975: A sociohistorical discourse analysis. *Language in Society, 25,* 352, 354, 358. Copyright © 1996 Cambridge University Press. Reprinted with the permission of Cambridge University Press.

Butler, Judith. (1993). *Bodies that matter: On the discursive limits of "sex."* New York: Routledge.

Keller, Evelyn Fox. (1983). *A feeling for the organism: The life and work of Barbara McClintock.* New York: W. H. Freeman. Copyright © 1983 by W. H. Freeman and Company. Used with permission.

Pinch, Trevor. (1994). Cold fusion and the sociology of scientific knowledge. *Technical Communication Quarterly, 3* (1), 85–100. Copyright © 1994 by the Association of Teachers of Technical Writing. Reprinted by permission.

Starr, Cecie. (1991). *Biology: Concepts and applications.* Belmont, CA: Wadsworth. Copyright © 1991 by Wadsworth, Inc. All rights reserved.

Taylor, Charles Alan. (1994). Science as cultural practice: A rhetorical perspective. *Technical Communication Quarterly, 3* (1), 67–81. Copyright © 1994 by the Association of Teachers of Technical Writing. Reprinted by permission.

I also wish to thank Kimberly Burgos, Acquisitions Editor for Ablex; Anne Trowbridge, Book Production Manager; and the Ablex production staff for their assistance. I am also grateful to Mike Markel for sharing his editorial expertise. I especially thank series editor M. Jimmie Killingsworth for his help and encouragement. And most especially to Marion.

NOTES

[1] For a useful survey of key works in the study of scientific discourse from literary theory and rhetoric to the history and sociology of science, see the introduction by Peter Dear (1991). For the uninitiated, David Locke (1992) offers an overview of the major issues in the study of scientific writing. Useful bibliographical tools include the sourcebook by Moran and Journet (1985) and the yearly bibliographies in *Technical Writing Teacher/Technical Communication Quarterly*. The first issue of the 1990 *Journal of Technical Writing and Communication* has a comprehensive annotated bibliography of its journal (Nakamura, 1990). See also chapters two and three of *Defining Science* by Charles Alan Taylor (1996) for an overview of the major contributions to the literature of science by 20th-century philosophers, sociologists, and historians of science. His extensive bibliography is also useful.

[2] Charles Bazerman (1983) summarizes the major theoretical arguments in the sociology of science before the mid 1980s. See also Steven Shapin (1982).

[3] Gaonkar (1993, pp. 267–270) proposes three classifications for rhetorical studies of scientific discourse. One classification traces a continuum from external to internal studies of science; that is, from studies of popularizations of science to science in public policy, science versus nonscience, studies of the way premises are transferred from "hard" to "softer" sciences and, finally, studies of scientific controversies within specific scientific disciplines. Literature in the rhetoric of science may also be divided into either general essays (those defining and justifying rhetoric of science and situating it in science studies) and case studies (those which take a scientific controversy or specific piece of scientific discourse and rhetorically analyze it). Finally, rhetorical studies of science may be identified by their basic analytical strategies; that is, whether they focus on a) the communal character of science as rhetorically situated and addressed; b) the epistemic status of science as rhetorically reasoned and argumentative; or c) the discursive dimension as rhetorically inventive for constructing knowledge claims.

[4] See especially Colin Russell (1983) for an understanding of the influence of history upon science in Europe between 1700 and 1900. The study by Thomas Kuhn (1970) of changing scientific paradigms straddles history, sociology, and philosophy of science. Too numerous to mention in this brief review are numerous articles analyzing the discourse of such noted scientists as Galileo, Newton, Darwin, Watson and Crick, and Einstein. See footnote 1 for bibliographical sources. In *Novelties in the Heavens*, Jean Dietz Moss (1993) studies the use of rhetoric in the Copernican debate. A major premise of her study is the need to distinguish the historical role of dialectic in science from that of rhetoric. Her analysis focuses on the writings of Galileo, but also includes a number of other participants in the controversy, including astronomer Johannes Kepler, Royal Society member John Wilkins, and theologians of the Roman Catholic Church.

[5] See, for instance, Ludwik Fleck (1935/1979), Imre Lakatos (1978), Chaim Perelman and Lucie Olbrechts-Tyteca (1958/1969), Karl Popper (1965, 1968, 1983), and Stephen Toulmin (1953).

[6] See, for instance, Clifford Geertz (1995), Stephen Jay Gould (1981), Gerald Holton (1993, 1996), and John Ziman (1981, 1984). Scott Montgomery (1996) offers a geologist's insights into the interplay of science, language, and culture.

[7] For a variety of perspectives on popular scientific writing, see especially a volume of essays edited by Murdo William McRae (1993). See also Stephen Jay Gould (1991) for a defense of popular science writing from the viewpoint of a practicing scientist.

REFERENCES

Bazerman, C. (1983). Scientific writing as a social act: A review of the literature of the sociology of science. In P. V. Anderson, R. J. Brockmann, & C. R. Miller (Eds.), *New essays in technical and scientific communication: Research, theory, practice* (pp.156–184). Farmingdale, NY: Baywood.

Bazerman, C. (1988). *Shaping written knowledge: The genre and activity of the experimental article in science*. Madison, WI: University of Wisconsin Press.

Bazerman, C., & Paradis, J. (Eds.). (1991). *Textual dynamics of the professions: Historical and contemporary studies of writing in professional communities*. Madison, WI: University of Wisconsin Press.

Dear, P. (Ed.). (1991). *The literary structure of scientific argument: Historical studies*. Philadelphia: University of Pennsylvania Press.

Fleck, L. (1979). *Genesis and development of a scientific fact* (F. Bradley & T. J. Trenn, Trans.). Chicago: University of Chicago Press. (Original work published 1935)

Gaonkar, D. P. (1993). The idea of rhetoric in the rhetoric of science. *Southern Communication Journal, 58,* 258–295.

Geertz, C. (1995). *After the fact: Two countries, four decades, one anthropologist*. Cambridge, MA: Harvard University Press.

Gould, S. J. (1981). *The mismeasure of man*. New York: W. W. Norton.

Gould, S. J. (1991). Prologue. In S. J. Gould, *Bully for brontosaurus: Reflections in natural history* (pp. 11–18). New York: W. W. Norton.

Gross, A. G. (1990). *The rhetoric of science*. Cambridge, MA: Harvard University Press.

Gross, A. G. (1993). What if we're not producing knowledge? Critical reflections on the rhetorical criticism of science. *Southern Communication Journal, 58,* 301–305.

Halliday, M. A. K., & Martin, J. R. (1993). *Writing science: Literacy and discursive power*. Pittsburgh, PA: University of Pittsburgh Press.

Holton, G. (1993). *Science and anti-science*. Cambridge, MA: Harvard University Press.

Holton, G. (1996). *Einstein, history, and other passions: The rebellion against science at the end of the twentieth century*. Reading, MA: Addison-Wesley.

Killingsworth, M. J., & Palmer, J. S. (1992). *Ecospeak: Rhetoric and environmental politics in America*. Carbondale, IL: Southern Illinois University Press.

Kuhn, T. S. (1970). *The structure of scientific revolutions* (2nd ed.). Chicago: University of Chicago Press.

Lakatos, I. (1978). *The methodology of scientific research programs*. London: Cambridge University Press.

Latour, B. (1987). *Science in action: How to follow scientists and engineers through society*. Cambridge, MA: Harvard University Press.

Latour, B., & Woolgar, S. (1979). *Laboratory life: The social construction of scientific facts*. Beverly Hills, CA: Sage.

Locke, D. (1992). *Science as writing*. New Haven, CT: Yale University Press.

McRae, M. W. (Ed.). (1993). *The literature of science: Perspectives on popular scientific writing*. Athens, GA: University of Georgia Press.

Merton, R. K. (1970). *Science, technology, and society in seventeenth-century England*. New York: H. Fertig.

Merton, R. K. (1973). *The sociology of science: Theoretical and empirical investigations*. Chicago: University of Chicago Press.

Merton, R. K., & Gaston, J. (Eds.). (1977). *The sociology of science in Europe*. Carbondale, IL: Southern Illinois University Press.

Montgomery, S. L. (1996). *The scientific voice*. New York: Guilford Press.

Moran, M. G., & Journet, D. (Eds.). (1985). *Research in technical communication: A bibliographical sourcebook*. Westport, CT: Greenwood Press.

Moss, J. D. (1993). *Novelties in the heavens: Rhetoric and science in the Copernican controversy*. Chicago: University of Chicago Press.

Mulkay, M. (1979). *Science and the sociology of knowledge*. London: G. Allen & Unwin.

Myers, G. (1990). *Writing biology*. Madison, WI: University of Wisconsin Press.

Nakamura, M. (1990). An annotated bibliography of *The Journal of Technical Writing and Communication*: 1971–1989. *Journal of Technical Writing and Communication, 20,* 45–107.

Nelson, J. S., Megill, A., & McCloskey, D. N. (Eds.). (1987). *The rhetoric of the human sciences: Language and argument in scholarship and public affairs*. Madison, WI: University of Wisconsin Press.

Pera, M. (1994). *The discourses of science*. Chicago: University of Chicago Press.

Pera, M., & Shea, W. R. (1991). *Persuading science: The art of scientific rhetoric*. Canton, MA: Science History Publications/U.S.A.

Perelman, C., & Olbrechts-Tyteca, L. (1969). *The new rhetoric: A treatise on argumentation* (J. Wilkinson & P. Weaver, Trans.). Notre Dame, IN: University of Notre Dame Press. (Original work published 1958)

Pickering, A. (Ed.). (1992). *Science as practice and culture*. Chicago: University of Chicago Press.

Popper, K. R. (1965). *Conjectures and refutations: The growth of scientific knowledge*. New York: Basic Books.

Popper, K. R. (1968). *The logic of scientific discovery* (2nd ed.). New York: Harper & Row.

Popper, K. R. (1983). *Realism and the aim of science*. Totowa, NJ: Rowman & Littlefield.

Prelli, L. J. (1989). *A rhetoric of science: Inventing scientific discourse*. Columbia, SC: University of South Carolina Press.

Prelli, L. J. (1993). Rhetorical perspective and the limits of critique. *Southern Communication Journal, 58,* 319–324.

Russell, C. (1983). *Science and social change in Britain and Europe: 1700–1900*. New York: St. Martin's Press.

Selzer, J. (Ed.). (1993). *Understanding scientific prose*. Madison, WI: University of Wisconsin Press.

Shapin, S. (1982). History of science and its sociological reconstructions. *History of Science, 20,* 157–211.

Taylor, C. A. (1996). *Defining science: A rhetoric of demarcation*. Madison, WI: University of Wisconsin Press.

Toulmin, S. (1953). *The philosophy of science: An introduction*. London: Hutchinson's.

Ziman, J. M. (1981). *Puzzles, problems, and enigmas: Occasional pieces on the human aspects of science*. Cambridge: Cambridge University Press.

Ziman, J. M. (1984). *An introduction to science studies: The philosophical and social aspects of science and technology*. Cambridge: Cambridge University Press.

part I

Rhetorical Interpretations of Science

1

Uncertain Science and the Sponsored-Research Process

Gay M. Gragson
Genetics Department
University of Georgia

Ted L. Gragson
Anthropology Department
University of Georgia

> The rhetoric of communication and the rationality of discovery are no longer separate, as they were before 1960; instead, they now share common ground.
> —Stephen Toulmin (1995, p. 4)

In this study, we analyze three molecular genetics grant proposals that were written by the same investigator and submitted to three different government funding agencies over a span of eight years. We place our analysis in the context of some of the current thinking (and some that has been around for a while) on philosophy of science and the ways in which such thinking is attempting to reconcile the changing nature of today's process of scientific inquiry with long-held ideas about the nature and method of science. The dialogue of the investigator's proposals and the responses from the grant reviewers demonstrates the repositioning taking place within the scientific community as its focus shifts from reductionism to whole-system approaches. Under the traditional view of science as an objective process of gathering facts to fill "black boxes," scientists determine the innate properties of entities by taking them out of the context of their network of relationships (Lloyd, 1994; Marsonet, 1995; Ulanowicz, 1986). Whole-system approaches to science, however, specifically seek to examine the relationships themselves. Thus, many modern scientists (particularly in the life sciences) are gathering the black boxes filled by previous investigators and reopening them to examine their internal relationships (Killingsworth & Palmer, 1992; Pickett, Kolasa & Jones, 1994). Yet, the search for Truth remains a firmly entrenched ideal of the scientific process. The result of the current repositioning is that the acceptable means of obtaining Truth are being broadened and deepened.

Perhaps the most significant catalysts for these changing perspectives are: a) the growing issue of scientific accountability—both ethical and fiscal, and b) the very rapid pace of technological development. Basic science, spurred on by concomitant technological innovations, has created a global paradox: At a time when the application of science has resulted worldwide in healthier and better lives, it is also proving to be a primary threat to those lives. Many of the environmental ills that we are now facing are potentially devastating, and finding solutions to them has become an urgent policy issue among governments worldwide. Increasingly, scientists are being asked to provide these solutions, and quickly. According to Funtowicz and Ravetz (1991), the task of modern scientists must be

> to address the pathologies of our industrial system by reacting to new problems and devising new methods. … We adopt the term "post-normal" [science] to mark the passing of an age when the norm for effective scientific practice could be a process of puzzle-solving in ignorance of the wider methodological, societal, and ethical issues raised by the activity and its results. … scientists now tackle problems introduced through policy issues where, typically, facts are uncertain, values in dispute, stakes high, and decisions urgent. (p. 138)

Urgency and uncertainty directly conflict with the carefully considered, conservative pace at which science has historically progressed and been communicated. Scientists have typically sought to achieve objectivity by referring to the world rather than by constructing a world. In this view of science, rhetoric was considered antithetical to the scientific objectives of clarity and thoroughness (Killingsworth & Palmer, 1992). However, today's problems frequently demand that scientists construct solutions without waiting to observe "the truth" (urgency). When clarity and thoroughness are hedged by urgency, not all the facts will be known about how and/or why the solutions work or what their ultimate effects may be (uncertainty). To complicate this scenario, rapid and highly sophisticated technological advances have, in fields like molecular genetics, revolutionized the kinds of questions that can be asked and answered. A dramatic example of this is the recent, successful cloning of a sheep. These factors have resulted in what Pera (1994, p. vii) has called the movement of science from "the kingdom of demonstration to the domain of argumentation."

What are the ramifications of this repositioning in terms of scientific rhetoric? A good starting point for exploring this question seemed to us to be the sponsored-research process. Generally speaking, the proposal review process, like the journal publication process, documents the exchange of ideas among scientific community members. It is, therefore, a likely point from which to view change or conflict. More importantly, the sponsored-research dialogue is critical because it determines which ideas will be allowed to become "science." As Myers (1990, p. 41) points out in his study of two scientists' writing, "The researchers must get the money in the first place if they are to publish [results] ... and participate in ... controversies." In this respect, the sponsored-research process provides a more current and accurate record of the struggle to define science than the reconstructed logic of peer-reviewed publications.

The cycle of events associated with the sponsored-research process—an agency-issued request for proposals, submission of a proposal generated by a scientific investigator(s), and the evaluation of the proposal by internal or external ad hoc panel reviewers—represents the rhetorical actualization of science. In this written discussion, scientific investigators voice their ideas for new science; agencies (which indirectly represent the public) voice the needs science can fill and pass final judgment; and reviewers (who represent the scientific community) voice the degree to which the investigators have met the stated need and/or the extent to which the investigators seem to be furthering the reviewers' definitions of good science within a given discipline. If the investigators are successful in their bid for funding, permission (in the form of a written letter granting access to funding) is extended to carry out those activities. At this point, the investigators' ideas have been agreed upon by all as constituting "science"—the results may be positive or negative, but it is agreed that they will add to scientific knowledge.

The three proposals we examined were submitted by the same investigator (whom we will call "the P.I.," for Principal Investigator) for essentially the same project to three different government agencies in 1987, 1991, and 1995, respectively. The first two proposals were collaborations between the P.I. and a colleague. The third proposal was submitted by the P.I. alone. We have placed our analysis against the backdrop of current scientific theory to help clarify the social and scientific issues that influence the course of research in the life sciences. In the following sections, we introduce the speakers in these exchanges and explain the scientific problem and proposed procedures, we provide a brief rhetorical analysis of each of the three proposals and their respective reviewer comments, and we finish with a discussion of current philosophy of science.

THE SPEAKERS

The dialogue of sponsored research involves three primary speakers: the investigators (scientists), the agency (source of funds), and the reviewers (usually peers of the investigators). Because the purpose of the agencies (see below) is to promote science, proposals that are funded qualify by definition as "science."

The Investigators

In general, the scientific investigators with the best chance of receiving funding from large government agencies will have a reliable record of productivity and be affiliated with an institution that has a reliable record of productivity. The P.I. is a tenured full professor in a genetics department and his collaborator is an associate professor in the school of forestry; both are faculty at a large land- and sea-grant university. The P.I. is a nationally recognized plant molecular biologist. His research laboratory has been continuously funded for other projects for nearly 20 years by agencies such as the National Institutes of Health, the Department of Energy, and the United States Department of Agriculture. We were initially intrigued, therefore, by the fact that the project discussed below has been repeatedly rejected for funding for nearly 10 years. The P.I. was able to make progress on this particular project through smaller grants from other sources.

The Agencies

Although there are many private and corporate sources of funding for science, in this study we deal only with government agencies. Arguably, government funding agencies as a whole constitute the most influential source of research funds in higher education in the United States today. At the university where the P.I. has his lab, for instance, 76 percent of all nonstate research funding in 1995 came from government agencies. Unlike private foundations, government agencies are politically and fiscally accountable to the tax-paying public. Consequently, the science that government agencies solicit and support reflects, to some degree, their response to public and political pressures. The agencies we discuss in this study are the United States Department of Agriculture (USDA), the Environmental Protection Agency (EPA), and the National Science Foundation (NSF).

The Reviewers

Agencies such as USDA and NSF enlist nonagency scientists to give expert evaluations of proposals and to recommend funding. Frequently, a proposal is given a preliminary review by an ad hoc group of subject-area specialists. This review is usually followed by a more formal panel review by disciplinary specialists who may have only general expertise in the subject or methods of the proposal. Ad hoc and panel reviewers alike are typically senior faculty members who have an established record of scientific achievements themselves. Agencies such as the EPA, however, rely on staff scientists, and proposals are not usually sent out for external review.

THE RESEARCH PROJECT

The P.I.'s proposed investigation was very dependent upon theoretical results achieved through technological means. The following summarizes the problem he sought to address and the means by which he proposed to do so.

The Scientific Problem

Rapidly increasing heavy metal contamination of soil, water, and air due to industrialization and urbanization is a global problem. The need to remediate such areas for public health reasons was acknowledged at the national policy level in the United States with the passage of the Resource Conservation and Remediation Act of 1984 (Smith, 1994). The remediation of polluted sites has been delegated to federal agencies, which then draw on fiscal resources such as those provided by the Superfund (managed by the EPA) to enlist the aid of the scientific community through the sponsored research process. In brief, the P.I.'s projects all proposed to genetically engineer plants (i.e., produce what are termed transgenic or chimeric plants) that can potentially remediate soil mercury (a toxic heavy metal) pollution. These transgenic plants would take mercury from the ground into their tissues via their root systems. Once in the tissues, they would either store the mercury as they store other elements, or process the mercury and expire it through their leaves as relatively harmless mercury gas. The research problem was how to engineer such plants, because mercury is toxic to most plants in even small amounts.

Mercury has industrial and agricultural uses that include bleaching paper and textiles, mining for gold, dressing seeds and bulbs to inhibit bacterial and fungal growth, and creating fungicidal sprays for fruit trees. Unlike organic pollutants that normally degrade over time, metal pollutants are immutable and require immobilization or directed relocation to control their effects. The rate of mercury release into the environment has decreased in recent years, but the metal continues to leach into wetlands, waterways, and estuaries from previously contaminated sites. The most dangerous mercury-based compound typically found at such contaminated sites is "organic" mercury, which is particularly toxic because it can concentrate 100-fold as it moves up the food chain. The consequences in animal and human populations can be tragic.

The Proposed Procedure

The P.I.'s interest in using plants to clean up heavy-metal waste in soil and water (called "phytoremediation") was initially sparked by the fact that there are naturally occurring, heavy-metal tolerant plants that inhabit areas with high concentrations of toxic metals such as mercury. The P.I. began with the basic science goal of exploring the genetic and biochemical nature of the hyperaccumulating systems used by plants to survive under these conditions by storing large amounts of toxins in their tissues. However, the P.I. speculated that it would be possible to alter more common or useful plants to contain similar capabilities, making them a very useful clean-up tool.

This innovative idea was based on the fact that certain kinds of bacteria that live in the soil defend themselves against some kinds of heavy metal ions by electrochemically reducing them to less toxic states (e.g., mercury salt [Hg^{2+}] to metallic mercury [Hg^{0}]). They possess an operon of genes that enables them to produce an ionic-mercury converting protein called "mercuric ion reductase" (merA). Other researchers had demonstrated in the laboratory that when the bacterial *merA* gene was transferred to the yeast *Saccharomyces cerevisiae,* it became weakly resistant to mercury (Rensing, 1992, cited in Rugh, Wilde, Stack, Thompson, Summer & Meagher, 1996).

This is the point at which the P.I.'s science, in 1987, left the realm of "referring" to the natural world and entered the realm of "constructing." Thus, it is also the point at which his puzzle-solving procedures departed from the norm. The lay person's version of the P.I.'s working hypothesis is as follows (C.L. Rugh, personal communication, August 29, 1996):

1. Bacteria, which are single-celled prokaryotes, possess a gene (*merA*) that enables them to process toxic mercury salts to less harmful mercury vapor;
2. These genes have been at least marginally successfully transferred to single-celled, eukaryotic yeast;
3. Therefore, it should be possible to successfully transfer the *merA* gene to multicellular, eukaryotic organisms such as plants, giving them the ability to reduce ionic mercury.

Initial attempts to simply transfer the bacterial *merA* gene to plants failed, and the P.I. decided to modify the gene sequence itself to be more plant-compatible. At about this time, a new, accelerated mutagenesis protocol using polymerase chain reaction (PCR) to change gene sequences became available. PCR technology was the bridge between the project as it was rhetorically expressed in the first proposal and the project as rational discovery because it enabled the P.I.'s laboratory to directly modify the *merA* gene to specifically fit their model plant. PCR technology constituted a major change in the rules for solving this sort of genetic puzzle because it enabled them to skip over many of the previously standard methods for achieving such a modification and to pinpoint relatively quickly the necessary gene modifications. What would have been a 10- to 12-year, staff-intensive project was accomplished (without major funding) in three years with a post-doctoral researcher and an undergraduate student.

To achieve the results that they have today, the investigators literally constructed a modified *merA* gene called *merApe9* using a modified PCR protocol to replace some 9 percent of the DNA coding region with nucleotide combinations and codons common to plants and bacterial *Escherichia coli* genes. The *merApe9* gene was eventually transferred into *Arabidopsis thaliana*, a small model plant with a rapid life cycle and simple genetics. These characteristics were important to the research because introducing the transgene and developing resistance in a plant is carried out according to Mendelian inheritance rules (standard genetic method), which means researchers must wait for the plant to produce seed before the stability of the gene transfer can be determined. Between 50–100 percent of seeds from *merApe9* transgenic *A. thaliana* lines germinated, grew, and flowered normally on mercury-containing agar at concentrations up to at least 20 ppm Hg^{2+}. The seeds, seedlings, or mature plants from nontransgenic controls or Mendelian segregants lacking the *merA* transgene did not germinate and/or died shortly after germination at concentrations between 5–20 ppm Hg^{2+} (Rugh et al., 1996). In their own words, the investigators envisioned the following application of their work:

> Selection of appropriate transgenic *merA*-expressing plants for each particular habitat could accelerate the biological transformation of bound Hg^{2+} to Hg^0 and its loss from these sites, over and above what is already processed by bacteria. The placement of these improved plants around pollution sources and at their points of discharge and collection could prevent toxic Hg^{2+} accumulation at and transport from these locations. ... Once plants have transformed the mercury, air movement will dilute it to nontoxic levels and remove the mercury from these areas. ... Ultimately, this should lead to a more natural distribution of mercury in the environment and lower mercury concentrations to nontoxic levels in those areas where it threatens wildlife and human populations. (Rugh et al., 1996, p. 3186)

The potential of using plants to clean their own environment excites the imagination of the lay person. In grant reviewers, however, it also excited a strong sense of caution, and even anxiety, as the following analysis shows.

PROPOSAL AND REVIEWER RESPONSE ANALYSIS

Proposal 1

The first proposal we considered was submitted in 1987 to the USDA Science and Education Office. It was written jointly by the P.I. and his collaborator in forestry. The grant only received funding for the portion the collaborator was to carry out. Their primary research goal was the "development of yellow poplar as a model plant for application of recombinant DNA technology to applied forestry problems" and their secondary goal was "the initiation of a new field of research focused on engineering useful heavy metal resistant plant species." Like the roles scientists create for themselves and their readers in their other professional writing (e.g., Selzer, 1993), the roles they create in proposals influence their audience's response (e.g., Myers, 1990). In this proposal, the P.I. and his co-author present themselves as unapologetically certain and enthusiastic about their ability to achieve their stated goals by using a construction rather than a discovery process. This tone is strongest in the 26-line Summary, where the investigators focus heavily on the creative aspects of their project: "latest in DNA transformation technology" and/or "technology" appear four times; "practical application," "useful," and "applied" appear a total of five times; and "engineering," "constructing," and "transforming" also are used on five occasions.

In the first paragraph of the Introduction, they claim: a) that "The proposed research will lead to the immediate development of mercury resistance" and b) that "This work will lead to the future development of many plant species resistant to mercury and other heavy metals which contaminate a large portion of the otherwise arable land in the world." In the second section, Justification and Significance, they claim that "creation and testing of … genetic transformation and tissue … regeneration are imperative" for forest scientists to be able to fully use molecular genetic technologies. They point out that although the proposal deals with mercury, "its utility in planning and developing resistance exists not only for mercury but for … all metals which have been reported to contaminate otherwise arable land." These are sweeping claims.

On the final page of this section, they devote a paragraph to the "Significance of heavy metal resistant yellow poplar." This paragraph begins conventionally enough with, "A commonly cited problem … is that there are very few single-gene traits of possible economic impact available to be transferred into [forest tree species] genomes." Sentence 2 picks up the applicability thread: "The gene for mercuric ion resistance is one that could have direct application to a real forestry problem." Sentences 3–6 link heavy-metal resistance to clean-up of contaminated land in the form of "hyper-accumulators"—plant tissues that accumulate otherwise toxic levels of heavy metals. They provide three citations for credibility. The paragraph ends with this visionary statement: "This approach [plant hyperaccumulators] could become a new method of processing low quality ore from mine waste and improve the ecological status of the land at the same time."

The review panel summary is gracious. It states that the proposal is "very interesting and worthy of support," but consists of "three separate projects." Significantly, the investigators were recognized as able to carry out the work (e.g., "[P.I.a] has excellent credentials … [P.I.b] has a very good tissue culture background…"). Nevertheless, the mercuric ion reductase gene development and insertion into yellow poplar segments of the proposal were rejected in favor of funding just the research into DNA transformation and regeneration (the first step of the three). The second and third steps received this neutral, inarguable reason for rejection: They "have no direct link to the forestry aspect of the USDA mission."

The individual reviewer comments, however, reveal a high level of discomfort with the P.I.'s rhetoric, the constructed nature of the work, and the overall degree of uncertainty surrounding it. Repeatedly their comments pick up on the P.I.'s broader visions for the work:

- I think the scientific value of attempting to introduce into plants whatever of the few genes that have been isolated and built into vectors is high enough, and does not require justification by the prospects of arboreal gold mining [referring to the use of trees to process low-quality ore];
- I am not persuaded that the engineering of mercuric resistance is wonderful. … Maybe it would have practical application … I don't believe they have a realistic view of what is involved. They make it sound like everything will work, and that it will be easy. It won't and it won't; and

- [the proposal] is a carefully prepared, thorough project ... however, ... the project is too ambitious and will take much longer than the three years proposed.

One reviewer was completely sidetracked by the uncertainty: "There was no mention as to how much acreage was impacted by heavy metals and as to how much of this acreage would ever be planted to yellow poplar."

The P.I.'s methods of pursuing the mercury resistance gene were also generally criticized, though vaguely. For example, their use of petunia and tobacco as model plants for perfecting the plant *merA* gene was criticized: "I question the need to work on petunia and tobacco ... why not develop the techniques on yellow poplar?" One reviewer was irritated by the broad scope of the proposed research, focusing criticism on the methods: "This proposal is cluttered with a lot of different types of approaches and experiments. As such, it suffers from a strong [sic; weak?] central focus."

Thus, although the USDA used as its final rejection the (politely) neutral fact that much of the project did not seem related to forestry, the reviewers were primarily put off by the strong, visionary language the P.I. used. The reviewers were perhaps justifiably concerned with the authors' emphasis on "look what we can do" and their inattention to "look what we can observe."

Proposal 2

The second proposal we considered, another collaboration between the same investigators, was submitted to the EPA in 1991. This proposal was rejected outright. It illustrates the potential for sharp division of purpose between policy that seeks to address urgency and reviewers who may have difficulty overcoming the inherent uncertainty in implementing that policy. The EPA solicitation for proposals specifically requested "exploratory research grants." In particular, the EPA wanted "investigations ... in these research disciplines which focus on any aspect of pollution identification, characterization, abatement or control, or address the effects of pollutants on human health or the environment." The call for proposals specifically stated that EPA did *not* want proposals that were "development of proven concepts." Clearly, the EPA was actively pursuing issues of urgency and uncertainty.

In this proposal, the investigators created a more cautious, fact-based role for themselves—less pioneering visionaries and more "traditional" researchers. The opening statement in the abstract is more to the point than that of the USDA proposal: "We propose a collaborative project, bringing together gene construction technology and tissue culture propagation to generate populations of transgenic trees of the forest species yellow-poplar, carrying and expressing genes capable of detoxifying heavy metals." They acknowledge the possible limitations of their work: "However, such a system could only be applied ... if ... Therefore. ..." Next, the authors try to anticipate their readers' questions: "(1) Will transgenic trees ... actually be capable of detoxifying metal ions and accumulating the reduced metal? and (2) Can plant gene promoters ... be applied to preferentially accumulate the reduced metal ions in leaves, stems, or roots?" They then describe two lines of research already underway (ethos) and generally outline their project goals.

This abstract is thus much more closely focused on the actual puzzle being solved than on their role in constructing it or the visionary potential of its application. They use passive voice constructions ("could be established" instead of "will"), and "technology" appears only once and "applied" only twice, and at that in a different context. In the Introduction, their vision of the applied nature of this research is limited to one conservative sentence: "We plan to make use of the latest advances in enzyme engineering and tree biotechnology to develop trees that can remove heavy metal ions from the soil and detoxify them." This is followed by "We have developed methods for genetic transformation …," and a detailed discussion of the enzymes involved, including a line figure showing the chemical equation of detoxified ionic mercury.

After briefly outlining their objectives, they provide a section on "Biological solutions to heavy metal contamination." However, in contrast to their first proposal, this section discusses only factual information about metal contamination of land and the genetic basis for mercury resistance. These pages are completely devoid of the speculative and visionary statements of the USDA proposal, although the potential applications are implicit and should be clear to an organization looking for solutions to soil pollution.

The remediation of heavy metal contaminated soil is particularly difficult. So-called in situ immobilization technologies (e.g., chemical leaching, vitrification) are prohibitively expensive because of the size of the areas affected and thus remediation typically consists of "increased mobilization of the metal ion" (Rugh et al., 1996). In lay person's terms this means removing the topsoil by scraping and then transporting it to an approved landfill. Thus, "remediation" of heavy metal sites as currently practiced is a misnomer since it actually consists of relocating the problem to a new site. This raises the specter of endangering the wildlife and ecosystems of new areas and further challenging the public trust in the decision-making ability of the executing agencies.

In view of these pressures on the EPA, this project would have seemed to be a shoo-in for funding. However, despite the agency's request for undeveloped research, the urgency could not offset the reviewers' distrust of uncertainty, which was reflected in their remarks. The first reviewer commented:

> This proposal would be excellent if more basic biological knowledge of heavy metals in plants were available. … The major flaw in this proposal is its feasibility … there is no guarantee the trees will take up enough mercuric ions to detoxify the soil. … Another problem is that if mercuric ions were taken up by the roots in large quantities and detoxified, where would the metal end up? … A risk is that if plants can be made to grow on such sites, why bother to enforce environmental standards preventing further contamination of such sites?

It seems to us that these criticisms overshoot the mark for a study that is only proposing to figure out how to put the *merA* gene into poplars. All of the issues this reviewer raises are concerned with problems of actual application, and they are questions that the P.I. could have no means of answering so far in advance of the basic research. This reviewer's language reveals him or her to be very uncomfortable, not so much with the research idea itself, but with the seeming lack of rules to limit the nature of acceptable solutions. This constitutes a clear example of what Funtowicz and Ravetz are calling post-normal science: The problem has been introduced by an agency (EPA) as a result of policy issues, and the solutions the policy requires will, by nature, have a high degree of factual uncertainty teamed with practical urgency. Yet the reviewer is simply not able to make that leap into the unknown. Despite the EPA's stated need for exploratory, undeveloped research in this area, the reviewer recommends not funding the work because at some point in time it might prove to be a too-effective solution.

The second reviewer is also uncomfortable that the investigators do not know how the mechanism will work before they have studied it: "[The proposal] fails in the area of whole plant physiology. The authors pose no mechanisms for the uptake and transport of mercury by poplar and how bacterial *mer* genes might facilitate these processes." Again, this comment seems to reveal a fussy discomfort that lacks focus—the poplars will uptake and transport the mercury the same way they do other substances, through their roots. How the *mer* gene might facilitate it cannot be known until the research is in progress.

Both reviewers also critique the authors' method of development, suggesting that "greater progress could be made if tobacco or petunia were used as the experimental plant to determine potential of this approach," a comment that directly contradicts the criticism received in the USDA grant. One also critiques the choice of yellow poplar ("Would another plant with a storage root such as carrot be more appropriate?") and the fact that the authors give no justification for its use. Suggesting the use of a food plant to store toxins stretches credibility. Another weakness the commentators found is that the paper does "not discuss problems with volatilization of mercury into the atmosphere once reduced. It seems more appropriate for a safe method of remediation to engineer the plants to only sequester the mercury" (i.e., keep moving mercury to different locations—status quo). The final "major weakness" the commentators found was the "uncertainty that the *merA* gene will function in plants." The concerns of these reviewers probably have some merit. Yet the futuristic nature of most of them seems strangely at odds with the specific agency request for exploratory research. Apparently, the reviewers could not recommend funding because the project did not satisfy, for them, a minimum level of certainty.

Proposal 3

The third proposal we considered was submitted by the P.I. alone to the NSF. The proposal was funded in the fall of 1995. In this proposal, the P.I. achieves an authoritative ethos through the data he has already gathered. The most noticeable feature of this proposal, at first glance, is that the first page includes a small but full-color photograph of engineered *A. thaliana* plants growing on mercury-polluted, soil-less medium (the USDA grant had no graphics on the first page; the EPA grant had a line graph). Immediately it is clear to any reader that the project has achieved substantial results, and the element of uncertainty is considerably diminished. The project summary begins with a statement about heavy metal contamination, and then describes the P.I.'s long-term goal, which is "to identify and utilize single gene traits to effect ecologically sound solutions to metal ion contamination based on phytoremediation." The P.I. spends a brief paragraph discussing what he has already found, then outlines his hypotheses, his specific aims, and ends the section with a sentence that outlines his vision of the application of this science: "In the future transgenic constructs will be examined in productive shrub and tree species, and fish for their potential use in the remediation of metal ion pollution." Compared to the first proposal, the P.I. considerably tempers his language about the utility of phytoremediation: "Hyperaccummulators do offer exciting potential solutions to the remediation of. … However. … Therefore, many years of research may be required to adapt. …" This grant also differs from the previous two in including a four-page "Work Completed" section.

The grant was awarded. Interestingly, however, the reviewer comments are not substantially different from those of the previous grants. The primary reviewer concerns seem to center on whether or not the research is basic (i.e., is it an acceptable Kuhnian puzzle?), the appropriateness of the methods being used, and (again) the uncertainty of application. Reviewer 1 says:

> The panel considered this as really two proposals. … The second proposal is considerably weaker than the first, partly due to a lack of detail, and partly because the panel could not gauge what basic knowledge this objective could provide. … a large part seemed to be a matter of applying the work, which would not be compatible with the scope of this NSF program.

Reviewer 2: "It was also not clear that even if the transgenic plants could be successfully transferred to field situations that the amount of mercury which could be removed from the soil would be of consequence." Reviewer 3 was more enthusiastic, basing support on the presence of convincing data: "They have clearly shown the growth capability … and detailed the expected activity of the enzyme" and "All of these studies form a very solid research foundation for this proposal given the existence of the current mutants."

Reviewer 4 also appreciated the data base: "[The P.I.] has a reputation for doing excellent work. . . . They have substantial data going into the proposal … The work to date has been outstanding." But he balks at the methods: "Unfortunately, the experiments are not well developed. How many plants will they use?" Of more concern: "I found this to be a wonderful example of an application for basic science … I didn't see that a basic hypothesis is being tested. This is really a very applied project [which] prevents me from rating this an outstanding proposal." Reviewer 5 picks up this thread: "It is the applied aspect of the proposed research that is also its major weakness. … it is premature to explore the alteration of the … protein." Reviewer 6: "[The P.I.] has an outstanding reputation … and has been a major contributor for some time," but "My only concern is whether a plant that removes ionic Hg from the soil but releases the volatile metallic form into the atmosphere will adequately address the problem. What is the fate of the volatile metallic Hg?" Reviewer 7 is also hung up on the application aspect: "Is it desirable to remove the mercury from the soil and put it in the atmosphere? I think not." Reviewer 8: "It is unclear what he will learn from these experiments. … Confidence in the success of the experiments would be significantly greater if preliminary data could be presented." Reviewer 9 is downright negative: "Why haven't mutant *merA*'s evolved naturally? To what extent have these been looked for? This is not discussed!" and "I'm not convinced that 'solid-to-gas' is an environmentally-sound phytoremediation strategy … [The P.I.] is a productive researcher, but not in this field and presents too little preliminary data to convince. … I'm not convinced of the arguments."

Reviewers 5 and 9, in particular, seem to be voicing concern at the constructed nature of the science the P.I. is proposing. "It is premature to explore" and "Why haven't *merA*'s evolved naturally?" hearken back to the fact that the P.I. is breaking with the traditional practice of referring to the world rather than constructing it. These scientists want the process described (referred to) as it naturally occurs (evolution, Mendelian segregation); they are not comfortable with having the process deliberately altered (constructed). Several reviewers seem to be equating the constructed aspect of the research with applied science. The research is not meeting their definition of basic science puzzle-solving, yet it can hardly be called applied because the P.I. is not proposing to engage in phytoremediation, just explore its mechanisms. In the end, however, although all reviewers had criticisms to offer, the fact that this proposal contains a solid base of data (the uncertainty is minimized, the puzzle is complete enough to reveal an acceptable form) seems to have been the deciding point in the decision to award the grant.

When we first proposed to review these grants, the P.I. and his laboratory had borne nearly a decade of official disinterest in their project, which seemed to them to have tremendous potential and implications. In the fall of 1995, with the granting of the National Science Foundation grant, the *merA* research was finally promoted to "scientific" status on a national scale. In April of 1996, the P.I. and his laboratory published their results in *The Proceedings of the National Academy of Science* (Rugh et al, 1996). In early April, the P.I. was approached by a *Wall Street Journal* reporter who wanted to do a feature article on phytoremediation. The P.I. agreed; the rest is history. On April 16, 1996, both the *Wall Street Journal* (Langreth, 1996) and the Science section of the *New York Times* (staff, 1996) featured articles on the *merA* phytoremediation project, and Paul Harvey featured it in his daily noontime radio editorial. Over the next few weeks, a number of articles appeared in various publications. In late spring, the P.I. was approached privately to create a company that will commercially provide phytoremediation technologies to those involved in site cleanup. In May 1996, the laboratory received an invitation to submit a proposal for further research on *merA* transgenes to a special program sponsored by the Department of Energy (DOE-SRS). Ironically, the program is designed specifically to move research ahead in the area of phytoremediation in order to begin remediation of industrial and military cleanup sites as soon as possible. In August 1996, the DOE awarded the laboratory another major grant that was submitted in early 1996. The study of phytoremediation is now "science."

It is clear from the three sets of proposal reviews, which vary surprisingly little, that as a group, the P.I.'s peers have been consistently skeptical as to whether this activity constitutes proper puzzle-solving science. He has been told, for instance, that his work is "technology" and not "science" (P.I., personal communication, September 18, 1995). But to use technology alone as the explanation for this schism (and in some areas it has become a schism) within the scientific community would be to miss an important point: Science as demonstration is a linear, logical enterprise following the physicalist Newtonian paradigm that is based on mathematics, logic, and physics (Pickett et al., 1994; Sober, 1993; Ulanowicz, 1986). The proposals we have considered challenge the Newtonian paradigm by seeking to unravel the complex relationships that constitute living systems, a notoriously difficult issue for physicalist definitions of science. While excellent at accounting for the function of mechanical (deterministic) systems, the physicalist Newtonian paradigm has limited usefulness in dealing with goal-directed (teleological) systems in which organisms act to further their own survival and reproduction (Pickett et al., 1994; Ulanowicz, 1986). The struggle within the scientific community to accommodate this schism between organic and mechanic systems seems to be approaching that of a Kuhnian paradigm shift (Kuhn, 1970). In the following pages, we review current thinking among some philosophers of science who in viewing science as a "semantic" process see a means for reconciling this gap.

SHIFTING PARADIGMS IN SCIENTIFIC DISCOURSE

Toulmin (1995) notes in his chapter in "Science, Reason, and Rhetoric" that philosophers of science have focused largely on formal aspects of the logic of method rather than on the rhetoric of science. Among scientists, however, the attempts to reconcile method and rhetoric are active and relate directly to the increasing attention environmental problems have received since the end of World War II. As early as 1956, at a conference entitled "Man's Role in Changing the Face of the Earth," Marston Bates stated in his summary remarks, "Perhaps our Western world … is starting a sort of process of suicide through its failure to assimilate science into its general culture." (Bates, 1956, pp. 1139–1140). At the same conference, Lewis Mumford spoke directly to the unacknowledged but undeniably rhetorical nature of science (1956):

> We must abandon the semantic hoax of reducing organic behavior to 'mechanisms,' for a machine is an arrangement of predetermined parts for serving a specific human end;

> and the conception of a machine reintroduces the very element of teleology that causal analysis attempts in vain to eliminate from the world of life. (pp. 1148–1149)

The central debate rests on whether a physicalist approach to science that seeks to understand organic systems by reducing them to a series of autonomous mechanisms can adequately accommodate the demands of remediating such systems when they are, in fact, dependent on networks of relationships. Science has historically achieved truth demonstratively through application of the methodological model of science. This model of science originated with early scientists (Bacon, Descartes, Newton, Whewell) and took its modern form in the 20th century under Popper, Lakatos, and Laudan. It is based on the following three theses (Pera, 1994, p. 4):

1. There is a universal and precise method that demarcates science from any other intellectual discipline;
2. The rigorous application of this method guarantees the achievement of the aim of science;
3. If science possessed no method, it would not be a cognitive and rational endeavor.

Under this model of science, there are three rhetorical conditions that must be met if research is to be considered "science" or "objective": it must be free from instrumental action that seeks to control the state of things; from policy action that seeks to change the state of things; and from poetical action that seeks to supplement the state of things (Killingsworth & Palmer, 1992, p. 106).

As Pera points out (1994, pp. 28–29), the paradox of using scientific method as a definition of science itself is that "science is characterized by scientific method, but a precise characterization of scientific method destroys science" because the more precisely method is configured, the more gaps appear in its ability to explain all the phenomena it is meant to explain. Any workable scientific method must account for the broad areas of ignorance that will constitute our knowledge about systems (see Funtowicz & Ravetz, 1991, p. 146). The adherents to the new positivism (neopositivists) forming the Vienna Circle in the early 20th century advanced the position that all true knowledge is empirical and based on observable data. Scientists were to use language to ascertain the truth of sentences concerning the world, and build theories which would thus be either verifiable or falsifiable. Unfortunately, the result was linguistic absolutism, in which scientific language could consist only of statements containing either analytic (*a priori*) or synthetic (*a posteriori*) propositions. *A priori* propositions are informed by logic and mathematics, while *a posteriori* propositions are informed by the empirical sciences, physics being the predominant discipline (Lloyd, 1994; Marsonet, 1995). The scientific theories that have been produced under this philosophy can be described as "syntactic," because such theories link statements about the world to facts about the world in a manner comparable to how rule-based grammar leads to syntactically structured language.

Syntactic scientific theories (the "received view" of science) are adequate for discussing mechanical systems (e.g., physics) because the relationships between mechanical entities are fixed and immutable, and researchers need only discover what they are. In most physical systems, the final state is determined by inital conditions. In living systems, however, the relationships between and among entities are neither fixed nor immutable. Such systems display what is called equifinal behavior because their final state may be reached from different initial conditions by following different paths (Bertalanffy, 1950). In the following example, the inadequacy of syntactic approaches to developing scientific theory for living systems is well-illustrated.

Karl Popper objected to Darwinian theory as tautological:

> The difficulty is that Darwinism and natural selection, though extremely important, explains evolution by "the survival of the fittest." Yet there does not seem to be much difference, if any, between the assertion "those that survive are the fittest" and the tautology "those that survive are those that survive." (as cited in Lloyd, 1994, p. 20)

Viewed in this way, it is easy to understand Popper's objections to Darwinian theory. In fact, this argument is currently enjoying renewed visibility in the heated political debates over creationism and evolution.

In response to the philosophical dead-end articulated by Popper, some biologists, rather than formulating additional syntactic theories built on analytic and synthetic propositions, have chosen to formulate what are termed "semantic" theories. These theories are built on families of models representing empirical phenomena (Lloyd, 1994; Parker & Smith, 1990). Rather than linking statements to facts via grammar, semantic theories rely on inherently meaningful models—akin to semantic units of language—that are independent from the phenomena and entities they may serve to represent. Again using evolution as an example, semantic theories "better accommodate the fact that evolutionary theory is bound to change as a result of the evolutionary process itself" (Lloyd, 1994, p. 18). Popper's apparent tautology disappears in the semantic version of evolution as viewed through Beckner's description of natural selection:

> Selection theory is a family of related models that explain or quasi-explain empirical generalizations and particular facts of evolution. The way in which selection theory effects a unification is to be sought in the relations between the models that are applied over the whole range of questions that the fact of evolution raises, e.g., questions about the origin of species and of the taxa of higher rank, the rates of evolutionary change, the development of adaptations, etc. (Beckner, cited in Lloyd, 1994, p. 21)

Beckner concludes that it is simply inappropriate to search for universal laws in biology equivalent to those sought in Newtonian physics. This conclusion is parallel to Pera's search for local rules. What evolutionary explanations/models do best is demonstrate that a certain (evolutionary) result is possible given the principle of natural selection and certain empirical assumptions.

The preeminent evolutionary biologist Richard Lewontin (1963) confirms this, emphasizing that theories consist of structures, and that these structures are constructed, not discovered. A "theory" is different in content and intent from other rhetorical scientific structures; it has empirical content, making it falsifiable/verifiable, and it is intended to explain something. Models, on the other hand, have no empirical content, but are entirely analytical. They can be used to make predictions and can themselves be tested against empirical information. A model, by itself, is neither true nor false in an empirical sense, although it can lead to the discovery of new phenomena by suggesting general results contained in but not apparent from collections of observations. Models thus give direction to future inquiry (Lloyd, 1994; Parker & Smith, 1990). Bertalanffy (1950) and Marsonet (1995) take this line of thinking and apply it to science at large by arguing that scientists who followed the procedures used in physics incorrectly assumed that biological entities could be understood totally independent of all relationships between them. Entities, however, "are merely nodes in a net of relations" (Rorty cited in Marsonet, 1995, p. 124).

As a molecular geneticist, the P.I. has successfully used a syntactic approach to his research, defining genes and gene function at ever-increasing levels of specificity with the help of advanced technology. However, the P.I.'s conception of the *merA* gene as a phytoremediator required a semantic approach in order for him to envision how the network of genetic and physical relationships might reasonably be expected to interact. As the analyses have revealed, peer reviewers have reacted strongly and quite negatively to the semantic aspects of his work. Yet, the NSF award suggests that agencies and reviewers are becoming cautiously acceptant of a semantic approach to the biological sciences.

CONCLUSION

The received view of science says that the practice of science is reducible to method and that understanding derives from physicalist statements. Pera (1994, p.vii) (and others) argue that science must be broadened from this "kingdom of demonstration" to include "the domain of argumentation" because science is as much a cognitive process as it is a practice-based enterprise (Kuhn, 1970). Scientific discourse must be persuasive, and is thus inseparable from the logos, ethos, and pathos of rhetoric. But it is rhetorical in a constitutive rather than ornamental fashion. Scientific claims are accepted only if they persuade the community within which they are put forward, through the exchange of arguments and counter arguments. Scientists may reasonably fear that in shifting the arbiter of the scientific process from method to the practitioners themselves, objectivity will be replaced with solidarity and rationality replaced by routine conversation. This need not be the case, however, if faith is placed in human reason and its capacities, and the correctness of statements is established in a rigorous manner by intersubjective discourse (Dolby, 1996).

When compared to methodological (syntactic) rationality, dialectical (semantic) rationality is a) ethically more tolerant because it is neither linked to a single property nor to a set of previously established requisites, but rather to a free debate over different properties and requisites; b) it is more adequate than an "anything goes" rationality because it does not depend on the whims of authorities or on external social factors and thus reflects actual scientific practice; and c) it is philosophically more attractive because it depends on the strength of arguments (Dolby, 1996; Pera, 1994).

In this study, we have only touched on the many and complex issues surrounding the rhetoric of science and the rationality of discovery. The conflict between "applied" science and "basic" science, for instance, has been substantially exacerbated by all of the changes we have mentioned in this study. This issue particularly affects academic institutions, which traditionally have been the domain of basic science. Objectivity is another issue. Within the granting process itself, for instance, the tone of the reviewer remarks reveal that true objectivity and dispassion remain elusive. Furthermore, it is common knowledge among scientists, particularly in highly competitive fields such as genetics, that editors and reviewers will hold manuscripts or reject proposals in order to further their own goals (for an interesting study on editorial subjectivity, see Hull, 1988). The confluence of urgency, uncertainty, and technology is challenging the tried-and-true at every level in science. Clearly, the common ground now shared by rhetoric and current, rational scientific discovery is an area of substantial importance for those who study the rhetoric of science.

ACKNOWLEDGMENTS

Gaining access to all parts of the written dialogue that constitutes the sponsored-research process can be very difficult because it is carefully guarded to ensure anonymity and objectivity. We are, therefore, very grateful to the P.I. for enabling us to review at length the proposals and reviewer responses discussed herein. We also thank his laboratory staff for always being willing to answer our many questions. Special thanks to J. Selzer, J. Harwood, C. Rugh, and the external reviewers for their helpful comments.

REFERENCES

Bates, M. (1956). Summary remarks: Process. In W. L. J. Thomas (Ed.), *Man's role in changing the face of the earth* (pp. 1136–1140). Chicago: University of Chicago Press.

Bertalanffy, L. von. (1950). An outline of general system theory. *The British Journal for the Philosophy of Science, 1*, 134–165.

Dolby, R. G. A. (1996). *Uncertain knowledge: An image of science for a changing world.* Cambridge: Cambridge University Press.

Funtowicz, S. O., & Ravetz, J. R. (1991). A new scientific methodology for global environmental issues. In R. Costanza (Ed.), *Ecological economics: The science and management of sustainability* (pp. 137–152). New York: Columbia University Press.

Hull, D. L. (1988). *Science as a process: An evolutionary account of the social and conceptual development of science*. Chicago: University of Chicago Press.

Killingsworth, M. J., & Palmer, J. S. (1992). *Ecospeak: Rhetoric and environmental politics in America*. Carbondale, IL: Southern Illinois University Press.

Kuhn, T. S. (1970). *The structure of scientific revolutions*. Chicago: University of Chicago Press.

Langreth, R. (1996, April 16). Altered weeds eat mercury particles in lab experiments on toxic waste. *The Wall Street Journal*.

Lloyd, E. A. (1994). *The structure and confirmation of evolutionary theory*. Princeton, NJ: Princeton University Press.

Marsonet, M. (1995). *Science, reality, and language*. Albany, NY: State University of New York Press.

Mumford, L. (1956). Summary remarks: Prospect. In W. L. J. Thomas (Ed.), *Man's role in changing the face of the earth* (pp. 1141–1152). Chicago: University of Chicago Press.

Myers, G. (1990). *Writing biology: Texts in the social construction of scientific knowledge*. Madison, WI: University of Wisconsin Press.

Parker, G. A., & Smith, J. M. (1990). Optimality theory in evolutionary biology. *Nature, 348*, 27–33.

Pera, M. (1994). *The discourses of science* (Botsford, Clarissa, Trans.). Chicago: University of Chicago Press.

Pickett, S. T. A., Kolasa, J., & Jones, C. G. (1994). *Ecological understanding: The nature of theory and the theory of nature*. New York: Academic Press.

Rugh, C. L., Wilde, H. D., Stack, N. M., Thompson, D. M., Summer, S. O., & Meagher, R. B. (1996). Mercuric ion reduction and resistance in transgenic *Arabidopsis thaliana* plants expressing a modified bacterial *merA* gene. *Proceedings of the National Academy of Sciences, 93*, 3182–3187.

Selzer, J. (Ed.). (1993). *Understanding scientific prose*. Madison, WI: University of Wisconsin Press.

Smith, R. K. (1994). *Handbook of environmental analysis*. (2nd ed.). Schenectady, NY: Genium Publishing Corporation.

Sober, E. (1993). *Philosophy in biology*. Boulder, CO: Westview Press.

Staff. (1996, April 16). After genetic engineering, a weed gobbles up mercury. *The New York Times*.

Toulmin, S. (1995). Science and the many faces of rhetoric. In H. Krips, J. E. McGuire, & T. Melia (Eds.), *Science, reason, and rhetoric*. Pittsburgh: University of Pittsburgh.

Ulanowicz, R. E. (1986). *Growth and development: Ecosystems phenomology*. New York: Springer-Verlag.

2

Neuropsychology, Sex, and Rhetoric

Christine M. Skolnik
The Pennsylvania State University

> As in rodents, the female default form may develop in the presence of an XY genotype, if androgens are not present in early life. Human XX individuals may also be masculinized in the presence of early androgens, to the point of exhibiting male genitalia. Other data indicate that the nervous system and subsequent behavior are also affected.
> —Kimura & Hampson (1993, p. 386)

> Indeed, "sex" is a regulatory ideal, a forcible and differential materialization of bodies, that will produce its remainder, its outside, what one might call its "unconscious." This insistence that every formative movement requires and institutes its exclusions takes seriously the psychoanalytic vocabulary of both repression and foreclosure.
> —Butler (1993, p. 22)

What specific issues and types of evidence fuel debates over the role of gender and the category of sex in the social sciences and humanities? Are sex and gender differences genetically "programmed" or are environmental factors more significant? Can language change the material conditions of sex or is language conditional on materiality? And does the rhetoric of recent scientific investigations into the category of sex have any bearing on contemporary work in the humanities and on the politics of gender?

Among the most interesting sites of current scientific inquiry into gender difference is neuropsychological research into cognition. The widely published research of Doreen Kimura, for example, investigates sex differences in brain structure, cognitive function, and hormonal dynamics from a behaviorist approach. Kimura's work is based on a number of assumptions characteristic of research in her field; at the same time, however, it provokes a reconsideration of some such basic principles. Much of Kimura's work, for example, rests on an assumption of cognitive differences between the sexes and seems to be motivated by a desire to discover the biological bases for those differences. (An evolutionary paradigm of sexual dimorphism undergirds the assumption of such social and biological distinctions.) In this field, furthermore, given sexual categories are mobilized as norms in order to describe and define both gender's group and individual behavior, and these sexual norms are reinscribed by the methodological approaches of neuropsychological research. However, Kimura's publications also reveal various factors which complicate established conceptions of sexual difference. What may be termed "the fluid dynamics of hormones," for example, undermines conventional notions of sexual difference as a stable material condition.

Within the context of critical/rhetorical theory, the highly acclaimed work of Judith Butler also problematizes, or at least complicates, traditional notions of gender and sexuality, as well as essentialist assumptions grounding much feminist theory. At the same time Butler's work is sustained by various practices within the humanities which remain in thrall to identity politics and an ongoing practice of rethinking/retheorizing the body.[1] Among Butler's key concepts is a dyadic heterosexual or heterosexist matrix, a social "grid" which requires individuals to fit into one of two mutually exclusive sexual categories, and which establishes a compulsory order of sex, gender, and desire. In opposition to this matrix and the seemingly stable material conditions upon which it rests, Butler posits "materialization," a material and linguistic process (neither essentialist nor constructivist), through which bodies come to matter. Key to this process of materialization are the citational and performative characteristics of language which have the potential to either reinscribe or disrupt social configurations.[2] By reading the production and destabilization of the category of sex in Kimura's recent publications within the context of neuropsychological practice in general, and through the lens of Butler's critique of sexual difference, I intend to illustrate how Kimura's discursive practices reproduce, extend, and in some ways undermine our general conceptions of sex, language, and materiality—and how the work of the scientist bears on the work of the philosopher.[3]

In the introduction to her monograph *Neuromotor Mechanisms in Human Communication* Kimura (1993) situates her work within the field of neuropsychology by discussing language in behavioral terms and within a "comparative/evolutionary framework" (p. 3).[4] This section of her book also lays out her assumption of a "continuity of modern human communication systems with those of other primate species, as well as with early hominids" (p. 4). Her chapter "Sex differences in brain organization" adheres to this evolutionary paradigm as it suggests that differences between the sexes are related to the "intrahemispheric organization of function" and "dimorphism in human abilities with males generally having better abilities relating to distance information and females to interpersonal space." The concept of dimorphism, here, is based on an implicit evolutionary narrative of social organization according to the category of sex (Kimura, 1993, pp. 157–159). Such a narrative is cited in "Neural and Hormonal Mechanisms: mediating sex differences in cognition," a chapter in *Biological Approaches to the Study of Human Intelligence*, which supplements Kimura's work in *Neuromotor Mechanisms* by focusing on the role of *hormones* in cognition, and which she co-authored with local colleague, Elizabeth Hampson (Kimura & Hampson, 1993, p. 376).[5] Much of the explanatory power of Kimura's research, then, is based on an onto-teleological evolutionary metanarrative: Evolution explains the origin of cognitive differences between the sexes in terms of their functions or purposes within a paradigmatic social and environmental context. This metanarrative, however, seems tenuously connected to Kimura's behavioral research methodology, which examines individual performance on isolated, clinical tasks. Although her hypothesis that "human communication has, during the course of evolution, become intrinsically bound to the various motor programming systems that control the relevant musculature," explains her focus on behavioral mechanisms, the clinical tasks designed to produce these behaviors are far removed from the sort of complex environmentally and socially situated tasks described in the evolutionary narratives. A closer look at Kimura's methodology will shed some light on this relationship between evolutionary paradigms, neuromotor mechanisms, and cognitive behavior.

In *Neuromotor Mechanisms* sex differences are discussed chiefly in the context of the relationship between brain morphology and function. Morphology is described in *quantitative* terms: Certain regions of the brain are identified as generally more developed or relatively more developed in females, others in males (Kimura & Hampson, 1993, p. 139). And function is chiefly discussed in the context of aphasia: Neuropsychologists trace the loss of particular linguistic abilities to corresponding lesions of the brain (pp. 140, 141). Kimura has also tested various sorts of *spatial* abilities in aphasic patients. In "Neural and Hormonal Mechanisms" sexual differences in cognition are constructed through three sites of investigation: brain organization/morphology, cognitive function, and "the role of sex hormones in determining individual differences" (Kimura & Hampson, pp. 377–378). In general, measurement of cognitive function interprets competency or fitness in the context of fabricated, clinical tasks. Among Kimura's tests of spatial ability, for example, are matching line drawings of hands and gloves, and recognizing constructed blocks, "three-dimensional models of Shepard-Metzler-type designs," in a series of photographs (Kimura, 1993, pp. 142–143). As I have previously suggested, such tasks are tenuously connected to any broader definitions of life skills or abilities, detached from any social or environmental contexts, and seem constructed to reveal prescribed differences. What I want to draw attention to here is an interesting tension in Kimura's work between her belief that human neural mechanisms are a result of evolutionary pressures including the social configuration of "hunter–gatherer" societies, and her empirical research which studies human beings outside of their socio-environmental contexts. The onto-teleological narrative recognizes an intimate relationship between environment and physiology, while the empirical methodology virtually ignores environmental factors.

Another interesting tension in Kimura's work is that between individual behavior and behavioral *norms*. In "Neural and Hormonal Mechanisms" Kimura and Hampson (1993) explicitly state that "apart from the intrinsic interest … such research provides a systematic means of investigating *individual* variation in behavior" (p. 375, emphasis added). The text, however, reinscribes general categories of maleness and femaleness by citing established assumptions about cognitive differences:

> Human males and females are known to differ, on average in their respective intellectual strengths. Women in Western cultures tend to excel on certain verbal abilities, such as fluency, spelling, and early articulatory accuracy … small-amplitude motor skills, or those requiring motor coordination within personal space. … Men, in contrast, are better on certain spatial tests … motor skills requiring accurate targeting of distant stimuli … tests of mathematical reasoning. (Kimura & Hampson, p. 376)

Although Kimura and Hampson subsequently qualify these general claims by explaining that implicit to a notion of "average differences" is "great overlap between males and females in their performance on such tasks," their statistical proofs tend to reveal differences *between* the sexes rather than differences within same sex groups (pp. 375–377). One might say that Kimura's work produces a stochastic law of sexual difference in which individuals are identified with their sex, which is in turn identified with *probable* behavior.

In this sense Kimura's clinical research and publications may reproduce normative sexual categories, and the clinical differences out of which these categories are constructed may or may not have much bearing on individuals in other contexts. Toward the end of "Neural and Hormonal Mechanisms," Kimura and Hampson (1993) signal their own distrust of or discomfort with such categories by placing the adjectival terms "male" and "female" in quotation marks (p. 390). In general, Kimura makes few claims about the social ramifications of her work; her conclusions are extremely circumspect (Kimura, 1993, p. 159; Kimura & Hampson, pp. 393–394). Still the evaluation of behavior and ability in accordance with the given categories of sex prevails. And such research is subject to various interpretations and misinterpretations. Kimura's work, for example, is cited in a 1995 *Newsweek* article by senior writer Sharon Begley who connects scientific evidence to sexual stereotypes. "Females excel at language, like finding just the right words to make their husbands feel like worms," Begley writes "Males can't verbalize even one good excuse for stumbling home at 2 a.m." (Begley, 1995, p. 48). Kimura is also cited by Anne Moir and David Jessel in *Brain Sex: The Real Difference Between Men and Women*, a controversial trade text which begins and ends with the opinion that men and women could achieve greater happiness by understanding and accepting their essential differences (Moir & Jessel, 1991, pp. 8, 43, 44, 190, 191). Kimura's totality of research, however, may be seen as resisting facile conclusions about sexual difference, especially in the area of hormonal influences on cognition.

The research published in "Neural Hormonal Mechanisms" produces "substantial" evidence of a connection between sex hormones and cognitive variation (Kimura & Hampson, 1993, p. 377). It traces differences in cognitive function among same sex groups with different levels of exposure to androgens at different periods in development (pp. 385–387) and also investigates the effect of hormonal fluctuation in females throughout the menstrual cycle (pp. 388–393). More specifically, the text describes a number of experiments investigating the relationship between testosterone levels and spatial ability and concludes that, "Within the *normal* range of androgens, there appears to be a curvilinear relationship, with the optimal level below that of the most androgenic males, but above that of the average female" (p. 386, original emphasis). In other words, though spatial ability is assumed to be a "male" characteristic, there is no *direct* correlation between testosterone level and spatial ability in clinical tasks (pp. 386–387). Also these hormonal differences and cognitive abilities are not consistent over race (p. 388).

Another section of this work reports the effects on cognitive behavior of fluctuations in estrogen and progesterone levels throughout the menstrual cycle (Kimura & Hampson, 1993, p. 388). "Finding systematic variation in cognitive patterns across the cycle," Kimura and Hampson conclude, "would strengthen the suggestion that intellectual functioning is in part determined by hormonal factors" (p. 388). The authors also describe the results of tests given to postmenopausal women undergoing hormone replacement therapy (p. 388). In both sets of experiments they set tasks previously related to sex differences (spatial ability, perceptual ability, verbal fluency, articulation, manual speed/coordination, deductive reasoning) and, to a certain extent, these experiments support general preconceptions about hormone levels and cognitive differences between the sexes:

> The findings for both studies indicate that, in fact, variations in female sex hormones have different and generally opposite effects on tests which differentiate males and females. The articulatory, complex manual, verbal fluency and some perceptual speed tests, on which women are often superior to men, generally showed enhancement in the high estrogen phase, compared to the low. Tests of spatial ability showed the reverse effect, in that they were depressed for women seen first in the high estrogen phase compared to those seen in the low phase. (p. 389)

Also interesting, however, are Kimura and Hampsons's suggestions that "'male' abilities and 'female' abilities have a tradeoff relationship" and that "estrogen fluctuations could merely be altering a balance between them rather than affecting them directly" (p. 390). This hormonal approach to the study of cognitive difference is particularly provocative because it traces the results of developmental and cyclical changes within the body as a fluid, dynamic system, challenging traditional assumptions of sex as a given or stable category of difference.

In general Kimura's work describes a complex relationship between hormones, chromosomes, development, and individual experience that eschews straightforward causal or teleological narratives. Her work is certainly not as deterministic as that of behavioral geneticist Thomas J. Bouchard, Jr., for example, who claims that "genetic factors are antecedent explanatory factors" (Bouchard, 1993, p. 50), nor as uncomplicated as popular representations of work in this area.[6] However, though Kimura's work challenges psychological preconceptions and complicates sexual difference, it does performatively assert a stochastic "law" of sex. That is to say, her statistical evidence assumes the relevance of norms, probabilities, and means to the discourse of individual cognitive difference. But how exactly does a stochastic law bear on individuals and individual processes? How would Butler respond to this stochastic law of sexual difference? And are there points of intersection or accord between contemporary scientific and contemporary philosophical approaches to gender?

Some further exposition of Butler's work may be necessary before I can embark on such a comparison. Judith Butler's *Bodies that Matter: On the Discursive Limits of "Sex"* builds on a theoretical dislocation of sex/gender/desire which Butler achieved with great academic success in her earlier work *Gender Trouble*. In general, this project challenges and deconstructs heterosexist hegemonies and the sociopolitical consequences of a rigid heterosexual cultural matrix (Butler, 1990, pp. 17–19; Butler, 1993, pp. 11–12). *Bodies* also questions the limits of language and materiality in a broad philosophical sense and points beyond gender politics, as it questions feminism's monopoly on the sphere of exclusion (1993, p. 42). More specifically, both *Gender Trouble* and *Bodies that Matter* evaluate apparent material constraints on the discursive articulation of sex and the consequences of a sociocultural matrix of compulsory heterosexuality through a *particular set of criteria*. For example, Butler's critique of "the compulsory order of sex/gender/desire," which is clearly laid out in the first chapter of *Gender Trouble*, focuses on the issues of gender *identity* and *sexuality* in its claims that desire is not constrained by sex and gender, and that gender is not only a cultural construct but one that, in fact, destabilizes the category of sex (1990, pp. 6, 7). This analysis and this focus is further developed in *Bodies that Matter* with Butler's critique of Plato's *Timaeus*, and particularly in the section subtitled "Improper Entry: Protocols of Sexual Difference" (1993, pp. 36–55). In general Butler focuses on issues of *identity* and *desire* (following issues raised in her first book *Subjects of Desire*, 1987), privileging these specific loci of gender difference.

As I have already mentioned, Butler's *Bodies that Matter* also interrogates the relationship between language and materiality in the formation of gender identity and sexual desire. In her introduction to this volume, Butler (1993) critiques essentialist and constructivist absolutes, and she contests assumptions that materiality and language are in a position of radical alterity (p. 8). She emphasizes both the *linguistic* elements of "positing" a material world and the *material* conditions of language:

> The body posited as prior to the sign, is always *posited* or *signified* as *prior*. This signification produces as an effect of its own procedure the very body that it nevertheless and simultaneously claims to discover as that which *precedes* its own actions ... This is not to say that the materiality of bodies is simply and only a linguistic effect which is

> reducible to a set of signifiers. Such a distinction overlooks the materiality of the signifier itself. (p. 30, Original emphasis)

Butler, then, is no radical social constructivist. She goes to some trouble elsewhere in the text to *deny* that her arguments imply that language originates or causes materiality (pp. x, 10). At the same time, however, Butler is reluctant to concede established assumptions regarding sexual difference. She denies her critics assurances of "sexually differentiated parts, activities, capacities, hormonal and chromosomal differences that can be conceded without reference to 'construction'" (pp. 10, 11):

> To "concede" the undeniability of "sex" or its "materiality" is always to concede some version of "sex," some formation of "materiality." Is the discourse in and through which that concession occurs—and yes, that concession invariably does occur—not itself formative of the very phenomenon that it concedes? To claim that discourse is formative is not to claim that it originates, causes, or exhaustively composes that which it concedes; rather, it is to claim that there is no reference to a pure body which is not at the same time a further formation of that body. In this sense, the linguistic capacity to refer to sexed bodies is not denied, but the very meaning of "referentiality" is altered. In philosophical terms, the constative claim is always to some degree performative. (pp. 10, 11)

Butler's reluctance to concede the materiality of the body, here, can be read as a reluctance to cite or performatively reinscribe the heterosexual matrix that produces sexed bodies as a given category of identity.

Butler (1993) also posits a process of "materialization" in opposition to or *between* established notions of the material as a stable, limited surface and radical constructivist notions of materiality. Following an incisive critique of false constructivist assumptions and consequences, Butler (1993) proposes to replace these "conceptions of construction" with "*a process of materialization that stabilizes over time to produce the effect of boundary, fixity, and surface we call matter*" (p. 9, original emphasis). She also conceives this process in Foucaultian terms, "in relation to the productive and, indeed, materializing effects of regulatory power," and presses beyond any *particular* interpretation of sex to theorize the very "matter" of sex. "Through what regulatory norms is sex itself materialized?" Butler asks, "[a]nd how is it that treating the materiality of sex as a given presupposes and consolidates the normative conditions of its own emergence?" (pp. 9, 10).

Butler (1993) continues her critique of the materiality of sex through a redefinition of performativity as *citationality*. Performativity, she explains, is not a "singular 'act,'" but rather (as Derrida "makes clear"), a "derivative" process which gains its impetus from an already established authoritative code (pp. 12, 13). She subsequently applies this understanding of performative citationality to the production of sexual norms:

> The norm of sex takes hold to the extent that it is "cited" as such a norm, but it also derives its power through the citations that it compels. And how is it that we might

> read the "citing" of the norms of sex as the process of approximating or "identifying with" such norms? Further, to what extent within psychoanalysis is the sexed body secured through identificatory practices governed by regulatory schemas? [*sic*]. (p. 13)

Butler's stated *intention* in this work, then, can be understood in terms of performativity, and the deviation of a citational chain "toward a more possible future to expand the very meaning of what counts as a valued and valuable body in the world" (p. 22).

Thus, Butler (1993) questions the materiality of sex in order to deconstruct a dyadic heterosexual matrix that produces, not only gender, but materiality. And she posits this destabilization of materiality in opposition to a hasty reinscription of the materiality of sex into poststructuralist discourse (pp. 27, 28). Furthermore, she identifies citational practices as both formative and potentially disruptive of a heterosexual matrix.

An integral part of my critique here is to suggest that Butler's work is relevant to other loci of inquiry into the matter of sex (breaks through what may be seen as the hermetic seal of her philosophical rhetoric), by arguing that Kimura's work is illustrative of an operative "heterosexist matrix," at the same time that it tacitly supports a concept of "materialization" which challenges general assumptions about sexual identity. In other words, Kimura's research submits to Butler's critique in some instances, and powerfully aligns itself with her arguments in other instances. This characteristic of Kimura's work—its ability to derive radical findings from a fairly orthodox context—reflects Butler's own extraction of materialization from her rereading of the tradition of western philosophy, and points to the possibility of interdisciplinary dialogue even among such disparately disciplined fields as philosophy and neuropsychology.

Returning once more to Kimura's work, I'd like to address her *citation* of an evolutionary master narrative to account for sexual dimorphism. In "Neural and Hormonal Mechanisms" Kimura and Hampson reiterate this narrative with specific reference to "hunter–gatherer" societies:

> Many of these sex differences may be understood within the context of evolutionary pressures for sexual dimorphism in the hunter-gatherer society in which our brains developed (Daly & Wilson, 1983). In this milieu, men had chief responsibility for the hunting of both large and small game, which would select for accurate targeting ability … accurate navigational skills. … Women in such societies had the primary responsibility for care of small children, and preparation of food, clothing. … These activities would be especially demanding of motor skills executed within reach, that is, within personal space. (Kimura & Hampson, 1993, p. 376)

The authors then proceed to consider various other biological and environmental factors and, to a certain extent, admit but never escape the tautological implications of their evolutionary story. Kimura and Hampson (1993) challenge environmental explanations by arguing that "any environmental factors must operate within these biological frameworks, making it problematic to separate experiential influences from physiological ones" (p. 377). However, they do not subject their environmental argument to a similar challenge: the argument that biological factors may also operate within environmental frameworks. In the context of Kimura's research, then, an evolutionary paradigm discursively evolves, or produces, a particular understanding of sexual difference, and it may reveal as much about the disciplinary evolution of neuropsychology as it does about the evolution of an archetypal "male" or "female" human brain (Halpern, 1992, pp. 7–9).[7] In other words, Kimura's assumptions regarding human evolution and sexual dimorphism can be seen as illustrative and constitutive of a heterosexual matrix within neuropsychological practice in general, which explains the origin of dimorphism, the pragmatic functions or purposes which drive and shape dimorphic behavior, and the relationship between morphology and cognitive behavior, within the context of two, mutually exclusive sexual categories.

Apart from identificatory practices, the field of neuropsychology also mobilizes norms (grounded on statistical evidence), which can be interpreted as constitutive of a heterosexual matrix. As I have already suggested, psychologists are aware of the causal chasm between tables of averages or means, calculations of typical deviations, measures of statistical significance and individual behavior: Individuals are not assumed to inhabit norms (Halpern, 1992, pp. 12, 13; Kimura & Hampson, 1993, pp. 375, 376). Still the measurement and the analysis continues. Cognitive behavior is reduced, in such research, to performance on isolated, clinically fabricated tasks. And the stochastic laws generated by Kimura's work, along with other research in the field of neuropsychology, is interpreted by lay-experts such as Anne Moir and David Jessel, authors of *Brain Sex*, to have general implications for social organization: "Men and women could live more happily, understand and love each other better, organize the world to better effect, if we acknowledged our differences. We could then build our lives on the twin pillars of our distinct sexual identities" (Moir & Jessel, 1991, p. 8). Such stochastic laws of sexual difference, then, may become exclusive "normative ideals" of "regulatory power" in a Foucaultian sense, and such laws obviously do—to the chagrin, disappointment, or vexation of critical theorists and political activists (Butler, 1990, pp. 9, 10, 16–18).

Kimura's work does not, however, merely reinscribe a heterosexual matrix. Although her publications clearly illustrate what Butler's has described as a matrix of dyadic sexual difference and also some epistemological consequences of the gaps it produces, Kimura's production of such disjunctions or gaps also *diverts* the citational chain of neuropsychological assumptions about sex and cognition. The scientist's work proceeds from an assumption of readily identifiable sexual difference and the foundational categories of "male" and "female" (Kimura & Hampson, 1993, pp. 375–376). At the same time, however, her work investigates individuals whose morphological, hormonal, or genetic makeup contradicts these categories (Kimura & Hampson, pp. 385–386). Most revealing is the breakdown of identificatory language which occurs within these sites of sexual ambiguity. For example, though the psychological subjects of her research are usually identified as either male or female, certain morphologically ambiguous individuals are referred to only by their genotype. "As in rodents," Kimura writes, "the female 'default' form may develop in the presence of an XY genotype, if androgens are not present early in life." The phrase "XY genotype," thus, replaces the term "male" in cases of morphological ambiguity. Similarly the term "female" is replaced by the phrase "human XX individual" in the comment that "human XX individuals may also be masculinized in the presence of early androgens, to the point of exhibiting genitalia" (Kimura & Hampson, 1993, pp. 385, 386). It is an interesting effect that in reading these definitions, the ambiguously sexed individuals seem not only hard to define but somehow subhuman because they lack sexual identity. In accordance with Butler's critique, a dyadic sexual matrix does indeed produce a category of exclusion and perhaps even abjection here.

Yet this very gap, this failure to articulate a sexual identity, begins to divert the citational chain of a heterosexual matrix, as do the other findings of Kimura's research into the relationship between cognition and hormonal fluctuation.[8] As I have previously mentioned, Kimura and Hampson (1993) signal the development of their own uneasiness with the constitutive adjectives "male" and "female" by placing them in quotations marks during their discussion of hormonal effects (p. 390). This discursive move begins to break down or at least to call into question the identification between individual cognition, sex, and hormones which launches their research.

Butler cites a similar scenario in the "Concluding Unscientific Postscript" to her discussion of "Foucault, Herculine, and the Politics of Sexual Discontinuity" in *Gender Trouble*. Following a consideration of the journals of Herculine Barbin,[9] she discusses genetic research into sexual identity which claims to have found "the binary switch upon which hinge all sexually dimorphic characteristics," but which stumbles when confronted with ambiguously sexed subjects whose chromosomal makeup does not correspond to their sexual identity (Butler, 1990, pp. 106–107). Butler's analysis of this genetic research pertains equally to Kimura's research in the field of neuropsychology:

> One might argue that the discontinuities in these instances cannot be resolved through recourse to a single determinant and that sex, as a category that comprises a variety of elements, functions, and chromosomal and hormonal dimensions, no longer operates within the binary framework that we take for granted. (pp. 109–110)

The connections here are obvious. Certainly there is scientific evidence to suggest that our given binary categories of sexual identity are problematic. And certainly the evidence that behaviors previously considered directly related to a stable material condition of sex are subject to developmental hormonal variations, that "male" and "female" cognitive abilities might be in a *trade off* relationship related to cyclical hormonal fluctuations, that cognitive abilities once considered "male" are most developed in individuals that may be described as tending toward hormonal androgyny—all these findings undermine conventional notions of sex as an originary and stable material condition of identity. And surely there is an epistemological affinity here between a hormonal process of cognitive/sexual *becoming* and Butler's concept of "materialization."

Kimura's neuropsychological research and Butler's philosophical critique, though perhaps not generally comparable or compatible discursive practices, accord with one another in that they both challenge the disciplinary traditions from which they emerge by problematizing assumptions about sexual difference, and in that they posit similar sorts of redefinitions of the sexed body. Also, both areas of inquiry have ramifications for research in other disciplines. Butler's work gestures to findings in scientific disciplines, while Kimura's work implicitly supports Butler's philosophical position. Butler identifies a heterosexist matrix which can be seen as an underpinning of Kimura's work, and Kimura points to sexual difference as a dynamic process in line with Butler's idea of materialization. Most provocative and most promising, however, is the evidence that these two methodologically and ideologically distinct inquiries into sex come to similar conclusions about the ambiguity and fluidity of sexual identity. Though I would not suggest that the categories of "male" and "female" are empirically or philosophically obsolete, I would assert that these terms require considerable qualification and reconsideration. And, no doubt, we could be more cognizant of the role of such categories in our own discursive practices if we hope to divert the citational chain of our given sexual identities in the interest of greater individual freedom and social equity.

NOTES

[1] Butler's work draws on and influences a variety of disciplines within the humanities. She is generally considered a philosopher, though her appointment at Berkeley is in Rhetoric and Comparative Literature, and she is widely read in English departments and Women's Studies programs.

[2] The "performative" in this context refers not only to the linguistic concept discussed in J. L. Austin's *How To Do Things with Words* (1975), but also to citational practices: "Within speech act theory, a performative is the discursive practice that enacts or produces that which it names ... In a critical reformulation of the performative, Derrida makes clear that this power is not the function of an originating will, but is always derivative ... the norm of sex takes hold to the extent that it is 'cited' as such a norm" (Butler, 1993, p. 13).

[3] I'd like to acknowledge two academic sources that provoke and legitimate such questions: Jack Selzer's *Understanding Scientific Prose* (1993) and Richard Doyle's ongoing research into the rhetoric of genetic and cryogenic practice. Both Jack Selzer and Richard Doyle are faculty members in Rhetoric and English Studies at Penn State. Doyle's *On Beyond Living: Rhetorical Transformations of the Life Sciences* is forthcoming on

Stanford University Press.

[4] In the dedication and acknowledgements preceding the "text" of her recent monograph *Neuromotor Mechanism in Human Communication*, Kimura acknowledges the influence of Hugo Liepmann who "might well be considered the first *behavioral* neuroscientist" (Kimura, 1993).

[5] Although I have tried to generally acknowledge Hampson's role as a researcher and author of "Neural and Hormonal Mechanisms," I will focus on Kimura as a leading and widely published neuropsychologist.

[6] I am referring here to works previously mentioned: Sharon Begley's *Newsweek* article, "Gray Matters," and the Moir and Jessel book, *Brain Sex*.

[7] I might further suggest that an evolutionary paradigm operates here in the manner of a "black box," what Bruno Latour has described as "ready made science" which enables "science in the making" by radically circumscribing an area of controversy and complexity, and reducing its epistemological and/or discursive function to "input and output" (Latour, 1987, p. 3, 4). My reading also alludes to Donna Haraway's *Simians, Cyborgs, and Women: The Reinvention of Nature* (1991), and particularly her ethnographic investigation into primatology, "The Contest for Primate Nature: Daughters of Man-the-Hunter in the Field, 1960–80," in which Haraway illustrates the relationship between the sociocultural dynamics of an intellectual community and the findings of its research.

[8] I am suggesting here that the effects of Kimura's discursive practices are as much a finding of her research into the relationship between cognition and sexual identity as her explicitly stated data and analyses.

[9] Foucault published the journals of Herculine Barbin shortly after publishing his *History of Sexuality*. Barbin, a hermaphrodite, was "assigned the sex of 'female' at birth" but was "legally compelled to change h/er sex to 'male'" in her early twenties "after a series of confessions to doctors and priests" (Butler, 1990, p. 194).

REFERENCES

Austin, J. L. (1975). *How to do things with words*. Cambridge, MA: Harvard University Press.

Begley, S. (1995, March 27). Gray matters. *Newsweek*, 48–54.

Bouchard, T. J., Jr. (1993). The genetic architecture of human intelligence. In P. A. Vernon (Ed.), *Biological approaches to the study of human intelligence* (pp. 33–93). Norwood, NJ: Ablex.

Butler, J. (1987). *Subjects of desire: Hegelian reflections in twentieth-century France*. New York: Columbia University Press.

Butler, J. (1990). *Gender trouble: Feminism and the subversion of identity*. New York: Routledge.

Butler, J. (1993). *Bodies that matter: On the discursive limits of "sex."* New York: Routledge.

Halpern, D. F. (1992). *Sex differences in cognitive abilities* (2nd ed.). Hillsdale, NJ: Erlbaum.

Haraway, D. J. (1991). *Simians, cyborgs and women: The reinvention of nature*. New York: Routledge.

Kimura, D. (1993). *Neuromotor mechanisms in human communication* (Oxford Psychology Series No. 20). New York: Oxford University Press.

Kimura, D. & Hampson, E. (1993). Neural and hormonal mechanisms: Mediating sex difference in cognition. In P. A. Vernon (Ed.), *Biological approaches to the study of human intelligence* (pp. 375–397). Norwood, NJ: Ablex.

Latour, B. (1987). *Science in action*. Cambridge, MA: Harvard University Press.

Moir, A. & Jessel, D. (1991). *Brain sex: The real difference between men and women*. New York: Carol.

Selzer, J. (Ed.). (1993). *Understanding scientific prose*. Madison, WI: University of Wisconsin Press.

3

Plotting a Middle Ground: An Account of a Failed Science Story

Mary Rosner
University of Louisville

"When *I* use a word," Humpty Dumpty said..., "it means just what I choose it to mean—neither more nor less."
—Carroll (1872, p. 124)

Evelyn Fox Keller, one of the most persuasive and influential of those feminists who write about the gendered rhetoric and epistemology of science, has taught many of us to recognize in science those qualities that Western Culture has long associated with masculinity—hierarchy, control, mastery, objectivity. Part of her persuasiveness may come from her own scientific background, which enables her to write from the vantage point of an insider, possessing an insider's loyalty to and perception of science as an exceptional route to knowledge in our culture. Keller typically attempts to balance her belief in that privileged status of science with a commitment to a feminism that characteristically challenges privilege. Although she often performs this rhetorical balancing act persuasively, in writing about geneticist Barbara McClintock, a misstep seems to occur: The story Keller intends to tell is not the story some feminists read. In this chapter, I will review the version of McClintock presented in Keller's 1983 biography of the geneticist, explain how Keller addresses a "misreading" of McClintock shared by a number of feminists, and suggest a rhetoric that might have better served her purposes with this audience.

KELLER'S BIOGRAPHY

Struggles over what will count as rational accounts of the world are struggles over *how* to see. —Haraway (1988, p. 587)

Keller's biography of Barbara McClintock, *A Feeling for the Organism* (1983), tells the story of an extraordinary individual. McClintock, a classical geneticist who worked with maize, made two significant contributions during her long career. Not long after receiving her Ph.D. in Botany in 1927, she co-authored a groundbreaking article on chromosome behavior to demonstrate what geneticists had long assumed: "an actual physical crossing over of segments of the paired chromosomes in question, resulting in the inheritance of a chromosome that derives partly from one parent and partly from the other" (Keller, 1983, p. 55). The value of her work was recognized by the scientific community when in 1939 McClintock was elected Vice President of the Genetics Society of America; in 1944, she was elected a member of the National Academy of Sciences (one of the few women to be so honored); and in 1945, she became the President of the Genetics Society of America, "a position no woman had yet held" (Keller, 1983, p. 118).

Although this response to McClintock makes her sound like a hero of a familiar success story, Keller suggests that the geneticist's second major discovery had a different reception: It was largely ignored, misunderstood, or trivialized. Beginning in the mid-1940s and continuing for more than 30 years, the geneticist created a series of experiments that raised questions about the explanation put forward in support of the Watson/Crick model of DNA, where information in a cell flows one way: It comes from the DNA and is copied to the RNA, which then becomes a blueprint for proteins responsible for genetic traits (Keller, 1985, p. 169). This theory suggests that the flow of information is regular, stable and, therefore, predictable. In contrast, McClintock's theory of transposition offered an explanation of the unpredictable: genetic *instabilities* (see Rosner, 1994, pp. 480–482). Many scientists responded to McClintock's presentations of her new theory in the 1950s with silence, mumblings, and snickerings (Keller, 1983, p. 138). Transposition received serious attention only after more than 20 years, culminating in the Nobel Prize in 1983.

This story that Keller tells in *A Feeling for the Organism* depicts McClintock not only as a productive scientist but as an anomaly who lived her life isolated and without human support, often misunderstood and occasionally ridiculed. That isolation is attributed in part to the science McClintock practiced. Pushing against the constraints of her discipline, questioning some of the tacit assumptions of her field, she was "an outsider" (Keller, 1983, p. xii). Rather than distance herself from what she studied, McClintock practiced an *intimate* science—one based on a total absorption in and partnership with nature—so that "the objects of her study [became for her] subjects in their own right"; they claimed "a kind of attention that most of us experience only in relation to other persons" (Keller, 1983, p. 200). McClintock also believed that reason alone was not enough to reveal the complexity and mystery of nature: sympathetic understanding, intuition, and mysticism—"'a feeling for the organism'" (Keller, 1983, p.198)—were essential too.

Through describing McClintock's deviance and difficulties, Keller's biography tells a story about scientific growth, which usually results from an interaction "between individual creativity and communal validation":

> But sometimes that interaction miscarries, and an estrangement occurs between individual and community. Usually, in such a case, the scientist loses credibility. But should that not happen, or, even better, should it happen and then be reversed, we have a special opportunity to understand the meaning of dissent in science. (Keller, 1983, p. xii)

In Keller's version, the community from which McClintock was estranged finally embraces her; her awards are evidence of its plurality; her story is the story of its growth. It is this message that Keller apparently wants to foreground as "more complex and more illuminating" than a reading of McClintock's life as a personal "tale of dedication rewarded after years of neglect—of prejudice or indifference eventually routed by courage and truth" (1983, p. xii).

While many readers accepted Keller's biography uncritically, some have raised questions about her story. It's been called overstated: "There is a strong implication in Dr. Keller's account that McClintock's ideas ... were all correct, and that her sceptical or uncomprehending colleagues were blinded by their devotion to successful but limited paradigms. ... This is disputable, at least" (Fincham, 1983, p. 378). It's been described as too highly colored by McClintock's own disappointment of "the lack of understanding she met when reporting her later discoveries" (Glass, 1984, p. 601). It's been labeled inaccurate, inadequate, oversimplified, idealized, essentialist:

- a misleading portrayal of "a mythic figure. ... a noble loner, whose vision and genius were ignored by her colleagues until very recently" (Koblitz, 1987, p. 401);
- one of those "idealized biographies ... [that] can minimize and obscure the real issues which lie behind one woman's life and work" (Grobicki, 1987, p. 211);
- a representative of not what science can be but of what it once was, "a return to a view of science in which the role of the scientist is reduced to reading off meanings written into the fabric of nature" (Dugdale, 1988, p. 124–125; see also Haack, 1992, p. 18);
- a faulty interpretation that contradicts testimony by McClintock's friends and colleagues (McGrayne, 1993, p. 172; also Glass, 1984, p. 601); and
- an example that runs counter to what Keller claims: she argues for plurality in science but clearly privileges McClintock's approach (Longino, 1990, pp. 207–208; also see Keller, 1989c, p. 723).

What Keller acknowledged as being particularly disturbing were not these criticisms but the actions of some feminists who refused to read the biography as a salute to science in spite of her explicit directions to do so. Instead, they connected

McClintock to a feminist science. Although in *A Feeling for the Organism* Keller professes to serve several groups and purposes—to introduce an unfamiliar world to readers who are not trained biologists; to "flesh out" a name from a textbook to students of classical genetics; to show professional biologists how discourse shapes the growth of science; and to help Keller herself "understand the relations between creativity and validation, between individual and community, and between one community's conception of science and another's" (1983, p. xv)—she apparently could not control against this particular reading of McClintock. Like Humpty Dumpty, she was left questioning who would be master of her words.

FEMINIST SCIENCE

> Language is not a neutral medium that passes freely and easily into the private property of the speaker's intentions; it is populated—overpopulated—with the intentions of others.
>
> —Bakhtin (1981, p. 294)

The feminist science that McClintock is said to exemplify is a suggested replacement for the stereotypically patriarchal science that has dominated the discipline, with its celebration of analytic thought, its claim to objectivity in order "to free us from distortion, bias, and error in intellectual inquiry" (Hawkesworth, 1994, p. 152); its aim "to 'attack' problems, and its goal ... to 'conquer' or 'master' nature" (Keller, 1974, p. 19). While it has not been defined in any detail, in this feminist science

> no rigid boundary separates the subject of knowledge (the knower) and the natural object of that knowledge ... the subject/object split is not used to legitimize the domination of nature ... nature itself is conceptualized as active rather than passive, a dynamic and complex totality requiring human cooperation and understanding rather than a dead mechanism, requiring only manipulation and control. (Fee, 1986, p. 47)

Keller's McClintock seems to practice a kind of science informed by these characteristics: She values understanding rather than mastery, kinship between self and other, intimacy, respect for complexity. But interpreting McClintock as a representative of a feminist science is in direct contrast with Keller's intentions.

Keller tried to make her intentions explicit both before and after the publication of *A Feeling for the Organism*. For instance, in "Feminist Critique of Science," published in the "Polemics" section of the British social sciences journal *Fundamenta Scientiae* in 1980, Keller argues that feminist criticism of science should "liberate, not ... reject, the scientific spirit" (p. 346). Thus, she cautions feminists against defining "feminine" in opposition to the rational and objective, since this definition both reinforces patriarchal oppositions which exile women to powerlessness outside of a culturally powerful science and makes objectivity solely a masculine goal. Those who use this oppositional definition fail to make available ways of being that have been denied to women, ways defined as "masculine."

This essay allows Barbara McClintock to be an example of an alternative science *only* if she is viewed carelessly: "A romantic view of 'woman' might … lead one to conclude that the science McClintock is engaged in is a radically different venture than 'men's science'" (Keller, 1980, p. 345). But such an interpretation, Keller warns, would be unfair to science and to McClintock: Equating McClintock's science with "'women's science'" means depriving "the rest of science of the opportunity, and indeed the necessity, of learning from her" (1980, p. 346). The correct way to see McClintock is as an exemplar of productive science, defined both by the "feminine" qualities of McClintock's work that are available to all creative people, including male scientists, and "by a period of long, arduous, highly purposive and disciplined investigation" (1980, p. 346). Thus, McClintock, who has joined intuition to "empirical and logical demonstrations of the utmost rigor" (1980, p. 345) with great success, transcends masculine/feminine categories. She is not "a feminist scientist" but an example of what Keller calls "a more hermaphroditic, or rather, genderless, conception of science than the one we are taught" (1980, p. 345).

Continuing her argument in "A World of Difference" from her collection *Reflections on Gender and Science* (1985), Keller again acknowledges that she "chose … not to read the story of McClintock's career as a romance" (1985, p. 161), not to see her science as a feminist science, and not to argue the "simple view" that more women with values like McClintock's in science would result in a new vision for science. Such a view, Keller advises, forgets that people who want to be scientists must become part of the scientific community; new members of a community are unlikely to cause any sudden changes to it. Making the geneticist an exemplar of a feminist science also ignores McClintock's own desire to transcend gender. This reading also wrongly suggests that McClintock's views of science were not shared—an interpretation that would mean that she had no place within the community of scientists, that she was no scientist. Instead Keller argues both that McClintock was not a complete outsider because science is more pluralistic than commonly acknowledged, and that McClintock's sex meant that she was not a complete insider either: While she was a scientist among scientists, she was also a woman in a world of men. To make a place for herself, McClintock redefined "gendered" concepts of mind and Nature. In "A World of Difference" (Keller, 1985) then, the geneticist represents the possibility of a "thoroughgoing transformation" of the discipline that would begin with scientists' rejecting gendered stereotypes.

In "The Gender/Science System," first published in 1987 in the feminist journal *Hypatia* and reprinted in Tuana's *Feminism and Science* (1989a), Keller again addresses continued misreadings of her McClintock. Here she begins by tracing the origin of the notion of a feminist science. At one time, Keller explains, women wanting to enter the field tried to repudiate their difference, their gender, to become part of an essentially patriarchal community. But with the recognition of multiple goals and standards in science, some feminists argued that difference was to be valued, not denied; for some, the value of difference led to the valuing of a different science, a *feminist* science. Keller argues that feminists could make McClintock an exemplar of a feminist science *only* by distorting McClintock, "rewriting. . . the text" (1989a, p. 37). They do this in two ways:

1. by changing difference to duality—into masculine versus feminine science—to deny the special status of the discipline, and
2. by transforming a "deviant and reclusive cytogeneticist" into a cultural hero "who, through her 'feeling for the organism,' seemed to restore feminine values to science" (1989a, pp. 36–37).

In this version of McClintock, the geneticist is "one of the girls," with "girls" being significantly different from "boys" in how and what they know. This McClintock exists in a world of "us" and "them," with the "us" able to offer novel theories (as evidenced by the Nobel Prize) by using qualities associated with the familiar feminine stereotype. To undercut this story, which makes McClintock representative of a different science, Keller focuses on the roles it assigns the geneticist:

- Implying that an exemplar of a science called "feminist" must see herself as a feminist, Keller rejects McClintock's connection to a *feminist* science since McClintock denied being a feminist.
- Keller also rejects the connection of McClintock to a *feminine* science. After all, McClintock disavowed feminine stereotypes, and she lacked the usual signs of "feminine socialization": for instance, she was never pregnant. Moreover, the notion of a *feminine* science ignores both the socializing process that transforms individuals into scientists and the historic oppositions between the concepts "feminine" and "scientific."
- Finally, Keller rejects the connection of McClintock to a *female* science, arguing that she was too iconoclastic to represent women and that men practice her kind of science as well.

While feminists seem to be the primary target in "The Gender/Science System," Keller also corrects an interpretation by mainstream science that places McClintock inside a monolithic view of science. This is the story of one big, happy, and productive science family told by scientists like Stephen Jay Gould, whose practices are said to be like the geneticist's in many ways. This story makes McClintock into "just one of the boys." Keller argues that this version, which denies the geneticist's difference, is unsatisfying because it offers no place for any significant diversity within science.

A third story suggested by Keller, one that she claims is not a distortion of McClintock, defines difference as multiplicity and offers—in place of a different science—difference within science. As a result, she redefines science as a house with many rooms with McClintock clearly in residence. This is the story attributed to those women scientists (like Keller) who resist the distortions of a patriarchal science and the notion of a feminist one. In supporting this interpretation, Keller appeals to the special status of science, claiming that an alternate science would "[undermine] confidence in the privileged attachment of science to nature" (1989a, p. 41). And she repeats arguments she used in earlier publications to reject *feminine* science: that associating the two yoked terms ("feminism" and "science") makes it an oxymoron and that claiming women necessarily practice a different kind of science "invokes the duality of sex" (1989a, p. 41) and asks women scientists "to give up their identity as scientists" (p. 42).

Keller traces the cause of these multiple stories about McClintock to politics and power ("the epistemic authority of scientists" [1989a, p. 40]): McClintock's difference is invoked or denied according to the value attached to science and to difference. Thus, mainstream scientists, with an investment in the special status of their discipline and much to lose if an alternate science were acknowledged, deny difference and see McClintock as one of their own. Since feminists have no necessary stake in the discipline, they can see McClintock as a symbol of another science. And women scientists who want to acknowledge both the special authority of science and the value of McClintock's difference place her within a canon that is broad enough to allow for her and flexible enough to learn from her. For Keller, this last interpretation identifies a "middle ground" that is "a clear advance over both the monolithic view of science that threatens to exclude diversity, and the dualistic ... view that threatens to deny [special] meaning to ... science" (1989a, p. 42). But she also notes that it ignores the role of gender to the story. For her, an even better "middle ground" would likely acknowledge that difference within science occurs in spite of cultural influences on the categories of gender and science that work against difference.

These days, McClintock gets little attention in Keller's work. In *Secrets of Life/Secrets of Death* (1992), for instance, the geneticist is cited briefly as an example of an alternative vision in science that has been persistent enough to survive but whose greatest success is insignificant when measured against the success of mainstream biology. No mention is made of "misreadings" although they persist in people like Dugdale (1988, p. 117), Fee (1988, p. 47), Wilshire (1989, p. 104), and Hekman (1990, p. 132). Keller's silence suggests defeat in her efforts to "master" her readers or "control" her text.

RHETORICS

> Not all metaphors are equally effective for the production of further knowledge.
> —Keller (1992, p. 33)

As I indicated earlier, Keller is committed both to a feminism devoted to nongendered conceptions of men, women, and science, and to a science that can encompass various methods and values, allowing her "to celebrate the differences which already exist within [it]" (Oliver, 1989, p. 141). She attempts to connect these two interests by arguing that feminist goals can best be met by uncovering gendered biases within science, not by embracing an alternative science inaccurately attributed to McClintock.

To make this argument convincing, Keller suggests time and again that her interpretation of McClintock is shaped only by the facts of the geneticist's life and the "truth" of science, not by any motive of her own. Thus, in "The Gender/Science System," for example, Keller explicitly denies any rhetorical intentions or manipulations behind her earlier biography of McClintock. Instead, she suggests that the version of McClintock she presented was not under her control at all: "It seemed unfair to McClintock's own story to burden it with a prior moral of mine, particularly in view of the fact that it was extremely unclear to me just what the relevance of a feminist, or gender, critique to that story might be" (1989a, p. 37). If gender was important, Keller continues, "the story itself would have to show both me and the reader how" (1989a, p. 37). In other words, the story told itself; the material spoke for itself. But Keller acknowledges that she was not an "innocent bystander" (1989a, p. 37): She enabled the story to be told, and she was interested in what it had to say. Although in all the ways that she defines herself as an interpreter of McClintock, Keller claims innocence of any rhetorical purposes or strategies; she is, of course, an accomplished rhetorician. She knows that "representations are never only matters of description ... they are also the stuff of politics. They reflect our tacit expectations and desires; they are stories we tell each other for the purpose of agonistic and other kinds of interpersonal interactions" (Keller, 1988, p. 247). Keller's positioning herself as a conduit for the "true meaning" of McClintock is itself a rhetorical act.

Keller is non-innocent in more ways than she confesses. In the texts discussed in this chapter, she repeatedly sets up oppositions between a good and a bad reading of McClintock and attempts to justify her "good" reading as the "true" and "only" one in several ways:

- by using an authoritative tone (made easy by her reputation in science and in feminism). This tone can both encourage belief and discourage dissent;
- by claiming the value of the position she espouses by suggesting that what is good for science is good for (all) feminists. This linking of science and feminists invites readers to see a place for feminism in science and suggests that rejecting one may mean rejecting the other; and
- by arguing that McClintock fails as the exemplar of feminist science however it is defined. By slipping back-and-forth among three terms—"feminist," "feminine," and "female"—Keller implies that McClintock has to be stereotypically female or feminine or a self-declared feminist to represent feminist science. Is this necessarily so?

Strategies like these reflect characteristics associated with conventional academic discourse.[1] A use of us/them oppositions, a hierarchy of interpretations with a single "right" reading and an "ideology of domination" (Longino, 1990, p. 206) reflect "'scientific' ways of knowing and valuing" (Shotter & Logan, 1988, p. 73) that Keller has long argued are gendered. As effective rhetoric, this kind of discourse "dominates other voices; it subdues the noise of other voices and is heard above them" (Alcorn, 1994, p. 20).

If the only way "to write well in Western culture is to write like a man" (Brody, 1993, p. 3), then Keller had no choice but to use a rhetoric that reflects some of what she herself condemns—to use the tools of the master to dismantle the master's house. After all, "to break with the conventions is to risk not being heard at all" (Tompkins, 1987, p. 171). But some feminists have labeled rhetoric like this as oppressive for "devalu[ing] alternative perspectives, understanding, and articulation of experience" (Lewis & Simon, 1986, p. 469). They have argued for a different kind of writing, a rhetoric that privileges complexities and ambiguities (Penelope & Wolfe, 1983, p. 126)—a feminist rhetoric. This kind of rhetoric values the multiple and the personal; views its audience with "equality, respect, and appreciation" (Foss & Griffin, 1995, p. 6); and asks for dialogue rather than debate, indirection rather than correction, and space to explore partial and multiple answers rather than a comfortable and—to use one of Keller's terms—*stable* closure. Feminist rhetoric can supplement the academic rhetoric that has proven so successful for Keller.

This alternative rhetoric might be especially effective when Keller is writing to feminists who are not necessarily allied to the values of patriarchy found in science and in academic argument. Persuading that kind of audience may take more than the old patterns of argument with its authoritative tone, linear development, and causal reasoning. It may take a rhetoric characterized by multiple meanings, risk-taking, and instabilities. To suggest how this alternate rhetoric might affect discourse, I would like first to examine the values underlying a metaphor that Keller uses in several of her essays, the "middle ground" metaphor, sometimes characterized as "a clear and stable 'middle ground'" (1989a, pp. 34–35). Thus, in *Secrets of Life/Secrets of Death*, she locates science in a "middle ground" between the extremes that bind it entirely to nature or to culture (1992, p. 6). One who has made a reputation by studying metaphors used explicitly and implicitly in science undoubtedly chose her own carefully: "Language is hardly free" (Keller, 1992, p. 6). We need to study Keller's metaphor carefully too. Following her lead—"feminists have accomplished the no mean feat of making visible the possibility of alternative mind-sets and metaphors" (1995, p. 90)—we can look at a different metaphor and an alternate "mind-set" that may more likely appeal to some of the readers she failed to persuade. In place of her "middle ground," I offer Susan Leigh Star's "high tension zone" (1991).

In arguing for "a clear and stable 'middle ground,'" for the desirability of fixed terms and balanced positions, Keller associates herself with stability, clarity, and even-handedness,[2] with balance and moderation as opposed to excess and uncertainty. Thus, in one text, Keller defines science so as to make herself a great mediator who avoids threats from what she describes as two extreme and destructive positions: "the Charybdis of 'scientific realism'" and "the Scylla of 'social relativism'" (1992, p. 9). Science is, she claims, something in between: It is a "middle ground" constrained both by a nature that is recalcitrant and by "social, cultural, and disciplinary location[s]" (p. 6) that are multiple. In another text, Keller criticizes features that can obscure the clarity and stability she considers essential for a "middle ground" account of a potentially volatile subject, features she characterizes as "ambiguity/unclarity/uncertainty/instability" (1989a, p. 35). She finds "instability" to be "an obstacle to productive exchange" (p. 44, note 1) and appears to champion a drive for fixed and single meanings.

This criticism of instability may be troubling for some feminists. Single fixed meanings often result from handy oversimplifications which can be powerful but silencing; witness the master molecule explanation of Watson and Crick, and the reception of McClintock's own alternative theory. Moreover, the attack on the mysterious and the ambiguous evokes and rejects images connected to another critic of science, Donna Haraway, who privileges "not the knowledges ruled by phallogocentrism (nostalgia for the presence of the one true Word) and disembodied vision. ... [but] those ruled by partial sight and limited voice" (1988, p. 590). Haraway would embrace instability, ambiguity, and multiplicity; the "stuttering, and the partly understood" (p. 589); and the need for continuing exchanges these would invite. In rejecting the partial and incomplete for the final and fixed, Keller may be alienating those feminists who prefer "the opening up of the text to ... new readings and new writings" (Engelstad, 1991, p. 507) and partial truths, those who do not necessarily want finality.

Keller's assumption of single and stable meanings also seems to suggest a positivist notion of rhetoric, one that reflects a view of words that Barthes describes as obedient and docile like a slave (Hayles, 1991, p. 5), and a view of readers who are also slave-like in being passive recipients of a meaning that cannot be questioned. These readers can understand that meaning as the author intended as long as the writing is "clear" and they are cooperative. Stability is a route to clarity, and clarity is universal and achievable. This notion is similar to the convention/fiction of the scientific report where facts seem to speak unambiguously and where "'messy' arguments" are not tolerated (Hudson & Jacot, 1991, p. 57). In science writing

> terminological ambiguity, uncertainty, and double entendre are generally seen as evidence of scientific inadequacy—as impediments simultaneously to progress and to truth and, accordingly, as impurities requiring removal. (Keller & Lloyd, 1992, p. 1)

Certainly, in writing about science, accuracy is essential, but accuracy does not come merely from words on a page; the notion that a writer can pin down meaning by the words he/she uses is a very old-fashioned view of how language works. These days we recognize that readers play an active role in making meaning based on what they bring to a text—for example, their knowledge of language, conventions, the world:

> what the reader brings to the work, what makes interpretation possible, is the familiar: the conventions of discourse, the world knowledge, the linguistic knowledge, the personal

> associations, that allow him[/her] to evoke a meaning from lifeless text. ... each reader will bring ... a different ... set of interpretive conventions, linguistic and world knowledge, and personal associations. (Brent, 1992, p. 31)

Because readers have different values, experiences, and backgrounds, words necessarily have multiple meanings, and different readers may read different texts. They "do not simply adopt the discourse systems they encounter; they admire, resist, or reject discourse according to their own unique character" (Alcorn, 1994, p. 14). Persuasion usually results, then, not by "freezing" an intended meaning and "numbing" a reader's mental activities but by helping the reader discover connections between the new story offered by the writer and the familiar story the reader already knows; that is, connections with

> the most general, deeply held, or important ideas [of the reader]. ... with values and attitudes that contribute to the organization of the reader's belief system ... and ... [with the reader's] entire structure of beliefs, attitudes and values. (Brent, 1992, p. 72)

Keller's "new story" about McClintock as a sign of the possibilities already inherent in patriarchal science may not have been shared by some feminists because it may not have connected with their already-existing attitudes critical of that science or offered them sufficient reasons to put those attitudes aside. As a result, Keller could only subdue, not persuade, these readers.

In contrast to Keller's "clear and stable 'middle ground,'" which evokes balance, comfort, and stasis, Star's "high tension zone" suggests disturbances, discomfort, and danger. Star uses this metaphor in "Power, Technology, and the Phenomenology of Conventions" (1991), which describes what happens to outsiders in a culture. She suggests that large organizations like McDonald's create stability by insisting on standardization. Although we are all members of "multiple social worlds" (p. 37) and, consequently, ambiguous/impure, standardization attempts to ignore differences in order to force us into a stable and single slot: "A set of uncertainties are translated into certainties: old identities discarded, and the focus of the world narrowed into a set of facts" (p. 47). Those who do not fit the standard—those, for example, allergic to onions—suffer inconvenience, alienation, pain; and "there are always misfits" (p. 36). In place of the stability that comes from the simplification of standardization, Star offers a "high tension zone," which results from a refusal to be slotted, to be pinned down. It is a state characterized by an uneasy ambiguity, "'as if something's about to explode'" (Star, 1991, p. 45). A "high tension zone" is "a kind of zero point between dichotomies ... or between great divides: male/female, society/technology, either/or" (p. 47). Entering a "high tension zone," Star suggests, may spotlight what is conventional or tacit for many, oppressive for some, leading to a fuller understanding, a *new* way of seeing.

In contrast to Keller's "clear and stable 'middle ground,'" then, which seems to confirm the status quo, Star's "high tension zone" values blurred boundaries, uncertainties, instabilities, potential danger, potential excess, and a potentially broadened perspective. Like Haraway, she recognizes that our multiple positioning forces a refusal of coherence, singleness, and stability because the cost may be too high.

MULTIPLE MEANINGS

> There is no context large enough to guarantee that a statement will have just one meaning, the intended meaning, that it will speak for itself. (Myers, 1990, p. 219)

Keller, like all of us, straddles multiple positions, including those of lapsed scientist and active feminist. And she often successfully joins these two positions for herself and for her audiences. But in the case of McClintock, Keller has apparently failed to turn some feminists to the "right path," and that failure may be attributed, in part, to the stance she takes. She allows only one reading of the geneticist; only her way is the right way, with McClintock as testimony for a single science, one that has "room ... for many of the questions, methodologies and interpretations that [the] more familiar version of [science] ... [does] not permit" (Keller, 1989a, p. 41). But Keller does not want to allow her own readers room for questions and interpretations because room might invite them to draw the "wrong" conclusions and lead them away from the story she wants to tell. As a result, intending, like Humpty Dumpty, to be master, she lectures, explains, declares, and demonstrates as she attempts to pin down a single reading of McClintock as a means to a broader view of science. Like the speaker in the poem Humpty Dumpty recites, "'[She] told them once, [she] told them twice: They would not listen to advice'" (Carroll, 1872, p. 132). In telling her readers just how to read McClintock again and again, she attempts to dominate them: She is the teacher, and they are the students to be force-fed; she is the actor, and they are to be acted on. Is it any wonder that at least some readers rebelled against this role? "'Of all the unsatisfactory—'" (Carroll, 1872, p. 136).

It may be that Keller could never have reconciled her multiple audiences to her vision of science, that her arguments were doomed to fail with some of that audience, and that McClintock was inevitably to be transformed in "inappropriate" ways. It may be that a feminist rhetoric using McClintock to argue the special status of science would have met with similar, only limited, success. But let me suggest that because it evokes open-ended exploration, feminist rhetoric might have made it more difficult for readers to slot McClintock "automatically into categories, ready-made places" (Code, 1995, p. 61) labeled "feminist science." And it might have discouraged Keller from offering her own master narrative of the geneticist. Fragmented and indirect, feminist rhetoric is not a panacea—"many readers may be left feeling a bit too destabilized" (Keller, 1995, p. 92); "its polyvocal, hybrid style of speaking might not affect the establishment at all" (Prins, 1995, p. 364)—but it is an option to a discourse that is patriarchal. If instead of insisting on a single reading that silenced others, Keller had employed this rhetoric to invite conversation, she might have told different true stories about McClintock and science, stories she could not own for readers she could not dominate.

As McKerrow reminds us, "Words do not move outward from the center in concentric circles. ... they are rocks skipped across turbulent waters, with each point of contact leaving its trace" (1993, p. 62). Perhaps they can be thrown at a single target with some accuracy and with some force. But Keller's targets and her texts were multiple and moving. She could do little to control either.

NOTES

[1] While the strategies used in the essays discussed in this chapter do not reflect the explicitly adversarial method of academic discourse that feminists like Frey (1990) and Tompkins (1987) talk about, Keller is clearly practiced in that method as well. She aggressively uses terms like "conflation" and "confusion" to attack Oliver (1989b, pp. 150–151). She accuses Richards and Schuster of suffering from a "failure ... to grasp," "deep misunderstanding," "misapprehension," and an implied simple-mindedness (1989c, pp. 721–722).

[2] But at least one article has commented on "the fundamental instability and manoeuvrability of that 'stable middle ground' to which Keller repeatedly lays claim" (Richards & Schuster, 1989, p. 725).

REFERENCES

Alcorn, Jr., M. A. (1994). Self-structure as a rhetorical device: Modern *ethos* and the divisiveness of the self. In J. S. Baumlin & T. Baumlin (Eds.), *Ethos: New essays in rhetorical and critical theory* (pp. 3–35). Dallas, TX: Southern Methodist University Press.

Bakhtin, M. M. (1981). *The dialogic imagination: Four essays* (M. Holquist, Ed., C. Emerson & M. Holquist, Trans.). Austin, TX: University of Texas Press.

Brent, D. (1992). *Reading as rhetorical invention: Knowledge, persuasion, and the teaching of research-based writing*. Urbana, IL: NCTE.

Brody, M. (1993). *Manly writing: Gender, rhetoric, and the rise of composition*. Carbondale, IL: Southern Illinois University Press.

Carroll, L. (1972). *Through the looking glass, and what Alice found there*. London: Macmillan.

Code, L. (1995). *Rhetorical spaces: Essays on gendered locations*. New York: Routledge.

Dugdale, A. (1988). Keller's degendered science. *Thesis Eleven, 21,* 117–128.

Engelstad, E. (1991). Images of power and contradiction: Feminist theory and post-processual archaeology. *Antiquity, 65,* 502–514.

Fee, E. (1988). Critiques of modern science: The relationship of feminism to other radical epistemologies. In R. Bleier (Ed.), *Feminist approaches to science* (pp. 42–56). New York: Pergamon.

Fincham, J. R. S. (1983). Genetics at first sight [Review of the book *A feeling for the organism: The life and work of Barbara McClintock*]. *Nature, 304,* 377–378.

Foss, S. K., & Griffin, C. L. (1995). Beyond persuasion: A proposal for an invitational rhetoric. *Communication Monograph, 62* (1), 1–18.

Frey, O. (1990). Beyond literary Darwinism: Women's voices and critical discourse. *College English, 52,* 507–526.

Glass, B. (1984) [Review of the book *A feeling for the organism: The life and work of Barbara McClintock*]. *Isis, 75,* 600–601.

Grobicki, A. (1987). Barbara McClintock: What price objectivity? In M. McNeil (Ed.), *Gender and expertise* (pp. 211–217). London: Free Association.

Haack, S. (1992). Science "from a feminist perspective." *Philosophy, 67,* 5–18.

Haraway, D. (1988). Situated knowledges: The science question in feminism and the privilege of partial perspective. *Feminist Studies, 14,* 575–599.

Hawkesworth, M. (1994). From objectivity to objectification: Feminist objections. In A. Megill (Ed.), *Rethinking objectivity* (pp. 151–177).
Durham, NC: Duke University Press.

Hayles, N. K. (1991). Introduction: Complex dynamics in literature and science. In N. K. Hayles (Ed.), *Chaos and order: Complex dynamics in literature and science* (pp. 1–33). Chicago: University of Chicago Press.

Hekman, S. J. (1990). *Gender and knowledge: Elements of a postmodern feminism.* Boston: Northeastern University Press.

Hudson L., & Jacot, B. (1991). *The way men think: Intellect, intimacy, and the erotic imagination.* New Haven, CT: Yale University Press.

Keller, E. F. (1974, October). Women in science: An analysis of a social problem. *Harvard Magazine,* 14–19.

Keller, E. F. (1980). Feminist critique of science: A forward or backward move? *Fundamenta Scientiae, 1,* 341–349.

Keller, E. F. (1983). *A feeling for the organism: The life and work of Barbara McClintock.* New York: Freeman.

Keller, E. F. (1985). A world of difference. In *Reflections on gender and science* (pp. 158–176). New Haven, CT: Yale University Press.

Keller, E. F. (1988). Feminist perspectives on science studies. *Science, Technology, & Human Values, 13* (3–4), 235–249.

Keller, E. F. (1989a). The gender/science system: Or, is sex to gender as nature is to science? In N. Tuana (Ed.), *Feminism and science* (pp. 33–44). Bloomington, IN: Indiana University Press.

Keller, E. F. (1989b). The gender/science system: Response to Kelly Oliver. *Hypatia, 3* (3), 149–152.

Keller, E. F. (1989c). Just what *is* so difficult about the concept of gender as a social category? (Response to Richards and Schuster). *Social Studies of Science, 19,* 721–724.

Keller, E. F. (1992). *Secrets of life/secrets of death: Essays on language, gender and science.* New York: Routledge.

Keller, E. F. (1995). The origin, history, and politics of the subject called "gender and science": A first person account. In S. Jasanoff, G. E. Markle, J. C. Peterson, & T. Pinch (Eds.), *Handbook of science and technical studies* (pp. 80–94). Thousand Oaks, CA: Sage.

Keller, E. F., & Lloyd, E. A. (1992). Introduction. In *Keywords in evolutionary biology* (pp. 1–6). Cambridge, MA: Harvard University Press.

Koblitz, A. (1987). A historian looks at gender and science. *International Journal of Science Education, 9* (3), 399–407.

Lewis, M., & Simon, R. (1986). A discourse not intended for her: Learning and teaching within patriarchy. *Harvard Educational Review, 56,* 457–472.

Longino, H. (1990). *Science as social knowledge: Values and objectivity in scientific inquiry*. Princeton, NJ: Princeton University Press.

McGrayne, S. B. (1993). Barbara McClintock. In *Nobel prize women in science: The lives, struggles, and momentous discoveries* (pp. 144–174). New York: Carol.

McKerrow, R. (1993). Critical rhetoric and the possibility of the subject. In I. Angus & L. Langsdorf (Eds.), *The critical turn: Rhetoric and philosophy in postmodern discourse* (pp. 51–67). Carbondale, IL: Southern Illinois University Press.

Myers, G. (1990). Narrative and interpretation in the *Sociobiology* controversy. In *Writing biology: Texts in the social construction of scientific knowledge* (pp. 193–246). Madison,WI: University of Wisconsin Press.

Oliver, K. (1989). Keller's gender/science system: Is the philosophy of science to science as science is to nature? *Hypatia, 3* (3), 137–148.

Penelope (S.) J., & Wolfe, S. J. (1983). Consciousness as style; Style as aesthetic. In B. Thorne, C. Kramarae, & N. Henley (Eds.), *Language, gender and society* (pp. 125–139). Cambridge, MA: Newbury.

Prins, B. (1995). The ethics of hybrid subjects: Feminist constructivism according to Donna Haraway. *Science, Technology, and Human Values, 20,* 352–367.

Richards, E., & Schuster, J. (1989). So what's not a social category? Or you can't have it both ways (reply to Keller). *Social Studies of Science, 19,* 725–729.

Rosner, M. (1994). Values in doing and writing science: The case of Barbara McClintock. *Journal of Advanced Composition, 14,* 475–494.

Shotter, J., & Logan, J. (1988). The pervasiveness of patriarchy: On finding a different voice. In M. M. Gergen (Ed.), *Feminist thought and the structure of knowledge* (pp. 69–86). New York: New York University Press.

Star, S. L. (1991). Power, technology and the phenomenology of conventions: On being allergic to onions. In J. Law (Ed.), *A sociology of monsters: Essays on power, technology, and domination* (pp. 26–56). London: Routledge.

Tompkins, J. (1987). Me and my shadow. *New Literary History, 19* (1), 161–178.

Wilshire, D. (1989). The uses of myth, image, and the female body in re-visioning knowledge. In A. Jaggar & S. Bordo (Ed.), *Gender/body/knowledge: Feminist reconstructions of being and knowing* (pp. 92–114). New Brunswick, NJ: Rutgers University Press.

4

Foucault's Archaeological Method and the Discourse of Science: Plotting Enunciative Fields

Cynthia R. Haller
North Carolina State University

In the years since Trevor Melia (1984) encouraged rhetoricians to turn their attention from ontological issues in the philosophy of science to the rhetorical analysis of scientific discourse (p. 312), rhetorical critics have directed their scholarship toward the examination of scientific texts (Campbell, 1986; Gross, 1990; Halloran, 1984; Lessl, 1985; Prelli, 1989, among others). Recently, however, Gaonkar (1993,1997a) has argued that such critical studies have revealed less about scientific discourse than about disciplinary politics—specifically, the desire of rhetorical critics to establish rhetoric as a metadiscourse. He further contends that classical rhetoric, since it was developed as a productive rather than interpretive framework, is "too thin" to serve as a useful critical tool for understanding scientific discourse. In particular, the authorial construct of the rhetor as an intentional strategist does not and cannot address the complex connections between social institutions and the production of discourse. A more useful critical framework for such purposes, Gaonkar claims, must be oriented less toward analysis of individuals' persuasive strategies and more toward analysis of the hegemonic desires embedded in discourse.

Gaonkar's argument seems to have hit a nerve with rhetorical theorists, rhetorical critics in general, and critics of scientific discourse alike. Responses to his position, which are numerous,[1] have critiqued and extended various aspects of his essay by exploring alternative possible definitions of the term "classical rhetoric"; the appropriateness of classical rhetoric as a critical framework; the relationship between interpretation and discourse production; and the role of author intentionality in discursive practice. I wish to take up some of the comments Gaonkar (1997b) has recently made about the last issue—specifically, to consider further his claim that the constructs of strategic persuasion and intentional agency are of limited usefulness in understanding the production of scientific discourse.

Gaonkar (1997b) argues that although specific individuals "deliberate, discuss, and make decisions," they do so "as points of intersection, relay, and circulation in a complex web of discursive practices" (p. 344). He asserts that the architecture of the web known as scientific discourse, because of its size and complexity, cannot be made apparent solely by a consideration of persuasive strategies used by rhetors in particular situations: "The vexatious fact remains that rhetoric, conceived primarily as a transaction by and between discrete individuals, cannot unlock the grammar of massive social formations such as 'modern science' that are propelled by 'system imperatives'" (pp. 337–338).

The paradigm of strategic persuasion may appear adequate for understanding scientific discourse, particularly when the definition of scientific discourse is restricted to mean only the discourse of scientists presenting research findings to other scientists. This type of situation accounts for much of the discourse that takes place within the pale of scientific communities; as Lyne and Howe (1986) have somewhat wryly put it, "Few scientists communicate with anyone outside of the small international fraternity that investigates the same vanishingly narrow specialty" (p. 141). In this type of discourse situation, however, the expected institutional function of a scientist is primarily persuasive. The goal is to expand the knowledge base of a scientific discipline through convincing one's colleagues of the validity and significance of certain findings (Lyne, 1990, p. 37); and the discourse that scientists generate in their roles as persuaders may be less a product of their own agency than a derivative of institutional discursive practices.

When the definition of scientific discourse is broadened to include not only science discourse (discourse written by and addressed to scientists) but also discourse about science (discourse about activities and findings of scientists but written by nonscientists for more public forums), anchoring discourse production in strategic persuasion becomes even more problematic. In discourse about science, objects of scientific research, scientific research activities, and even scientists themselves are characterized very differently from the way they are characterized in science discourse. Fahnestock (1986), for instance, has argued that science discourse is typically forensic, contains qualifications and hedges, and focuses on conjectural and definitional stases, while discourse about science is typically epideictic, implies the certainty of scientific findings, and tends toward qualitative and procedural stases (pp. 278–285). Killingsworth and Palmer (1992) have noted that science discourse presents scientific research activities in skeptical, open-ended narratives, while discourse about science presents them in mythologized narratives characterized by high drama (pp. 143–145). Lyne and Howe (1986) point out that paleontologist Stephen Jay Gould, in science discourse that he writes for other paleontologists, portrays himself as a Darwinist trying to reconstruct Darwinism, while creationists' discourse about science portrays him as an ally in the war against evolutionary theory (pp. 141–142).

It is certainly possible to account for these differences in the portrayal of research activity using traditional concepts of speaker, audience, purpose, and persuasive strategies as the relevant categories of critical dissection. The most commonly offered explanation for the differences is that a difference in audience necessitates a difference in strategy. Fahnestock (1986), for instance, attributes stasis differences between scientific discourse and discourse about science to a change in audience: Mass media publications must relate science to larger public issues to capture a public audience (p. 291). Similarly, Lyne and Howe (1986) suggest that Gould's discourse underwent a series of transformations as it moved steadily outward through a set of ever-widening audiences, the most narrow being the scientific discipline of paleontology, the next being other scientific disciplines such as biology and genetics, and the last being the widest audience of all: "Everyman" [sic], numbering in the "tens of millions" (p. 141). Indirectly, Killingsworth and Palmer (1992) assert the influence of audience on shaping mass media discourse when they speak of how journalists estimate the "information value" of various scientific research efforts: To attract public audiences, scientific research must be news (able to tell the public something it doesn't already know) and must be of human interest (applicable to problems already accepted as important issues by the general public) (p. 134).

As helpful as the vocabulary of classical rhetoric may be in accounting for variations between science discourse and discourse about science in specific situations, however, its focus on strategic persuasion within individual rhetorical transactions prevents it from illuminating fully the interconnections between the two types of discourse and the nature of the institutional power relations within which they are embedded. And yet it is precisely these connections and relations on which the grammar of both science discourse and discourse about science are predicated. Killingsworth and Palmer's (1992) and Lyne and Howe's (1986) interpretation of the differences between science discourse and discourse about science, for instance, both stretch naturally toward an institutional accounting of discourse. Lyne and Howe view creationists' appropriation of Gould's discourse as a move to use the authoritative status scientists generally hold in U.S. society as a warrant for their own position (p. 140). Killingsworth and Palmer suggest that scientists' tendency to distance themselves from public concerns in their discourse is a reflection of the scientific community's desire to establish itself as separate from (and by implication, superior to) the general public and its concerns (pp. 148–151). These interpretations directly attest to the fundamental role of broad social and institutional forces in the production of discourse.

For the analysis of such forces, Michel Foucault's archaeological method provides an alternative to more traditional rhetorical frameworks. Foucault's potential contribution to rhetorical criticism has already been established by Blair and Cooper (1987), Foss, Foss, and Trapp (1991), and Foss and Gill (1987). Blair and Cooper, for instance, have pointed out that Foucault's system for the analysis of discourse can reveal configurations of knowledge and power in society, as well as help the critic understand where and how discourse is changed or transformed (p. 162). Blair (1987) has argued that President Lyndon Johnson's statements in his San Antonio speech constitute acts of knowledge, power, and ethics that ultimately served to justify Johnson's approach to the Vietnam conflict (p. 375). Cooper (1988) has suggested that President Richard Nixon's Checkers speech is significant not so much because it illustrates the use of apology as a rhetorical strategy as that it demonstrates the kind of "knowledge, power, or ethical posture [that] is created when someone apologizes" (p. 12). The discourse Blair and Cooper have studied, however, has been political rather than scientific. The value of Foucault's methodology for understanding the relationship of science discourse to discourse about science has not been adequately explored. His methodology, which characterizes discourse not as a product of authorial intention but as a matrix within which relations of knowledge and power in society are created, maintained, and transformed, offers a way of ascertaining how science discourse and discourse about science are connected to one another within a web of power relations among scientific communities, the mass media, social institutions, and the general public.

Using Foucault's method to investigate these connections, I explored the configurations of power and knowledge revealed in two contemporaneous instances of scientific discourse, one an example of science discourse and the other an example of discourse about science. Both instances, or discursive events (Foucault's term for an occasion or situation where discourse is produced), ostensibly address the same topic: Chronic Fatigue Syndrome (CFS). Chronic Fatigue Syndrome has emerged as an object of discourse only recently, over the past 35 years, and its name transformations during those years suggest the multiplicity of its possible discursive constructions. Over 50 names have existed for the illness: Yuppie Flu, myalgic encephalomyelitis, Chronic Epstein-Barr Syndrome, Chronic Fatigue Immune Dysfunction Syndrome, and Chronic Fatigue Syndrome are but a small sampling of the names that have been used to identify it. Using Foucault's archaeological method, I analyzed and compared a speech delivered by Dr. David Bell, a prominent CFS researcher (Bell, 1990),[2] and a *Newsweek* feature article on CFS (Cowley, Hager & Joseph, 1990).

In general, both of these discursive events raise the status of CFS research and researchers within the medical community, but paradoxically they at the same time make contradictory statements about the medical community itself. Bell's speech, an example of science discourse, characterizes his work and CFS research in general as a logical step in the ongoing research agenda of a competent medical community, whereas the *Newsweek* article characterizes Bell's work and the work of other CFS researchers as an act of heroic rebellion against an indifferent medical community. Thus, Bell's speech legitimates and reproduces the authority of the medical community in society, while the *Newsweek* article undermines it.

Since an analysis of two single discursive events is not a sufficient means of characterizing entire discursive formations,[3] no broad claims about the nature of and differences between scientific and mass media discourse formations can be made from this study alone. The analysis does show, however, that Foucault's analytic categories are able to reveal some of the intersection points between discursive events, which in turn reveal some small part of the complex matrix of power relations known as modern science. This capability suggests that archaeological method may provide a useful tool for critics interested in exploring how the production of scientific discourse is embedded within institutional structures and discursive formations. Further, science discourse and discourse about science appear to be interconnected in complex and sometimes paradoxical ways; examining them as if they were two distinct discourse worlds, while seemingly warranted because of their obvious differences, may obscure some important complementary relations between them.

DISCOURSE, POWER, AND KNOWLEDGE

Foucault's conception of discourse differs from conceptions common in traditional rhetorical theory. Foucault developed his understanding of discourse outside the framework of rhetoric entirely, proposing it as an instrument for investigating the history of systems of thought (Foss, Foss & Trapp, 1991, p. 209). In *The Archaeology of Knowledge*, one of his most in-depth discussions of his archaeological methodology, he argues for a new understanding of how discourse should be used for historical research. Traditional historical methods, he claims, consider documents to be representative traces of the past through which a historian can discover the underlying truths, or monuments, of the past. To Foucault, however, the documents *are* the monuments (Foucault, 1972, pp. 6–7). The appropriate objects of inquiry for the historian are therefore documents themselves, not the events and continuities of ideas that have generally been assumed to exist somehow behind and apart from documents. The aim of the archaeologist is a "pure description of discursive events," in order to ascertain the regularities and unities that form within discourse itself.

Such regularities reveal power and knowledge relations in society. Power can be defined as "an open, more-or-less coordinated (in the event, no doubt, ill-coordinated) cluster of relations" (Foucault, 1980, p. 199); and knowledge goes hand-in-hand with power, since "a knowledge untainted by relations of power cannot exist" (as cited in Foss, Foss & Trapp, 1991, p. 205). Knowledge, therefore, does not develop in some uniform, progressive manner consistent with the discovery and representation of some underlying truth that exists apart from discourse; rather, it develops in a pattern consistent with the development of the relations of power to which it is connected, which are in turn consistent with the development of patterns in the production of discourse. Knowledge concepts do not preexist discourse but are inscribed within the totality of their discursive uses:

> The history of a concept is not wholly and entirely that of its progressive refinement, its continuously increasing rationality, its abstraction gradient, but that of its various fields of constitution and validity, that of its successive rules of use, that of the many theoretical contexts in which it developed and matured. (Foucault, 1972, p. 4)

Since the contextual uses of a concept involve discursive practices and discursive formations (collections of discursive practices), a "pure description of statements" enables the archaeologist to understand the relations of knowledge and power that exist in society. Indeed, paradoxically, discourse is the very means by which these relations are constituted (Foucault, 1972, p. 116). Ideology is thus neither prior to nor separate from discourse, but is, as Cooper (1989) puts it, "a knowledge/power product of rhetorical action" (p. 30).

The goal of uncovering the workings of knowledge and power through the analysis of discourse is ultimately an ethical one: to understand how human beings are transformed into subjects (Foucault, 1982, p. 208). Understanding how power turns human beings into subjects ultimately enables them to strategically free themselves from excesses of the process. Foucault points out this tactical aspect of archaeology in his introduction to *I, Pierre Riviere* (1975), an edited reproduction of a 19th-century dossier containing documents concerning the title figure's murder of his mother, brother, and sister:

> Documents like those in the Riviere case would provide material for thorough examination of the way in which a particular kind of knowledge (e.g. medicine, psychiatry, psychology) is formed and acts in relation to institutions and the roles prescribed in them (e.g. the law with respect to the expert, the accused, the criminally insane, and so on).
>
> They give us a key to the relations of power, domination, and conflict within which discourses emerge and function, and hence provide material for a potential analysis of discourse (even of scientific discourse) which may be both tactical and political, and therefore strategic. (p. xi)

But exactly how does an archaeologist go about achieving a "pure description of discursive events"? As Foucault begins to dissect his method in *The Archaeology of Knowledge*, he identifies the statement as the atomic unit of discourse. His definition of the statement, however, is elusive and complex, with several definitions being proposed and then withdrawn (Kusch, 1991, p. 60). Indeed, he seems inclined to define a statement by describing what it is *not* (Blair, 1987, p. 367): It is not the equivalent of a sentence, proposition, or speech act (Foucault, 1972, p. 86).[4] This negative definition of the term "statement," however, is perhaps not so much a scholarly failing as a telling reflection of the assumptions of archaeological method regarding the nature and significance of statements. What is of interest to the archaeologist is not so much the content or propositional value of a particular statement as the conditions under which it is formed. In other words, the archaeologist seeks an answer not to the question "What does this statement mean?" but rather "How is it that one particular statement appeared rather than the other?" (Foucault, 1972, p. 27). The answer to this more important question is what reveals the web of institutional power and knowledge relations within which particular statements come to be uttered.

The significance of a given statement, then, is discovered not by examining the statement in isolation, but by determining the characteristics of the entire enunciative field in which the statement was formed. Kusch (1991), taking literally Foucault's claim that dealing with statements means dealing with their enunciative function, has described this enunciative field mathematically as an operational field. Thus, the archaeologist achieves a pure description of statements by identifying the entire operational field in which the statements are located, which is composed of four subfields: "a field of objects ... a number of possible subjective positions ... a domain of coordination and coexistence [and] ... a space in which they [statements] are used and repeated" (Foucault, 1972, p. 106, as cited in Kusch, 1991, p. 60).[5] In keeping with Kusch's breakdown of the enunciative function, Blair and Cooper (1987) suggest that a "pure description of statements" involves determining the ways statements constitute their subject matter; the ways they constitute the role or position of the speaker; their conceptual relationship to other statements within a discursive event; and their comparative function or use with respect to other types of statements (p. 164). A statement's enunciative field can also be delineated by plotting the four coordinates in Foucault's series of discursive regularities: the formation of objects; the enunciative modalities or subject positions; style or ways of organizing concepts; and strategies (Kusch, 1991, p. 64). It is this series of coordinates that I used for my analysis.

Because a statement's significance is dependent on its unique location within an enunciative field constituted by multiple factors, no single iteration of a statement can be considered identical to another. The trademark of discourse is difference, division, and dispersion; and each time a statement is uttered in the context of a particular discursive event, the complexity of the statement's field increases. A statement thus manifests itself as an entity that possesses a "repeatable materiality." It is an object, and as such can reappear within a number of discursive contexts, but the character, identity, and nature of that object will be altered in accordance with the conditions of the various discursive events in which it is used (Foucault, 1972, p. 105).

The implication of repeatable materiality is that a given statement will be formed, used, appropriated, and reused in different contexts according to the desires of the institutional structures at work in society; and thus the archaeologist must pay particular attention to difference, contradiction, and disparity in statements and their uses (Foss, Foss & Trapp, 1991, p. 231). The aim of archaeology is to understand discursive practices as systems of dispersion (Foucault, 1972, pp. 37–38), since the patterns of contradiction and difference in discourse reveal configurations of power and knowledge in society—and one of archaeology's key attributes is its emphasis on examining, comparing, and understanding difference:

> Archaeological study is always in the plural; it operates in a great number of registers; it crosses interstices and gaps; it has its domain where unities are juxtaposed, separated,

> fix their crests, confront one another, and accentuate the white spaces between one another. (Foucault, 1972, p. xx)

To use archaeological method involves engaging in "comparative descriptions of discursive practices with each other and of discursive practices with the elements surrounding them" (Foss, Foss & Trapp, 1991, p. 201). In the following section, I will describe the differences I found in the enunciative fields of the two discursive events I analyzed, Bell's speech (1990) on Chronic Fatigue Syndrome (CFS) and a *Newsweek* article by Cowley, Hager and Joseph (1990) on the same topic, tracing how these differences locate the generation of discourse within institutions rather than individuals. The coordinates I used to plot the enunciative fields were the formation of objects, the formation of subject positions, the uses of style, and the uses of strategies.

CHRONIC FATIGUE SYNDROME: TWO PERSPECTIVES

Bell's speech, uttered by a physician to an audience composed largely of other physicians, formulates CFS as a phenomenon that can be isolated and subjected to the rigors of scientific research. Bell also inserts his research into an ongoing research tradition sanctioned by the medical community. He uses quantitative data, charts, and tables—hallmarks of the medical-scientific establishment—to support his arguments. Finally, he relies on themes of etiology and objectivity for his strategy.

Cowley, Hager and Joseph, on the other hand, formulate the disease as subjective experience: CFS may be a disease, but it is pertinent as a disease only by virtue of the way it affects the lives of human beings. Bell's research effort, in contrast to being presented as a part in the machine of the medical establishment, is portrayed as a mythic tale of rebellion and adventure. The graphics used in the *Newsweek* article, rather than contributing to an objectification of the illness, appeal directly to the reader's emotions. Finally, the authors rely on strategies of human subjectivity and theatricality.

The Formation of Objects

The first coordinate of the enunciative field is the formation of objects, or how a statement constitutes the object to which it refers. Understanding how an object becomes formed in discourse requires mapping three things: the surfaces of its emergence, its authorities of delimitation, and its grids of specification (Foucault, 1972, pp. 41–42). A surface of emergence is a discursive event where the object is accorded some status as an object (an object does not exist until someone speaks of it). The type of status ascribed to the object, however, is determined not by individual rhetorical choices but by the nature of the institutional site where it emerges. Foucault notes, for instance, that madness emerged variously as "disease, alienation, anomaly, dementia, neurosis or psychosis, degeneration, etc." within different social institutions such as medicine, religion, law, and the family (1972, p. 41). Bell's speech and the *Newsweek* article are both surfaces of emergence for CFS, but are situated within differing institutions—the speech within the institution of medicine and the article within the institution of public journalism.

Authorities of delimitation are those institutions that are commonly recognized as the ones with the right to name and define a particular object. In the case of CFS, the Centers for Disease Control (CDC) is one authority of delimitation; and CFS's validity as a disease is asserted in both Bell's speech and the *Newsweek* article by invoking the CDC criteria for diagnosis. The medical community as a whole, however, another major authority of delimitation for diseases, was at the time in some disagreement about whether CFS ought to be accorded the status of disease. Many doctors and medical authorities attributed CFS symptoms to depression, psychological illness, or some other medical condition. This lack of agreement within the medical community constituted part of the conditions of existence for Bell's speech. By the time of Bell's speech and the *Newsweek* article, the CDC had officially attributed disease status to CFS by formulating criteria for diagnosis, but many physicians remained skeptical of the disease, since no definitive etiology had been agreed on for it. The CDC's list of symptoms could as easily be signs of other illnesses, and the onset of CFS could not be reliably assigned to any particular viral or bacterial agent. Bell was invited to make his speech by an organization of Chronic Fatigue Syndrome sufferers in upstate New York with the intention that his speech might educate members of the local physicians' association, many of whom were skeptical of and/or uninformed about the disease. One goal of the organizers was thus to increase physician acceptance of CFS as a valid disease, and even though the speech was open to the public, a large portion of the audience consisted of medical professionals.

Both Bell's speech and the *Newsweek* article affirm the disease status of CFS, but an analysis of their grids of specification reveals significant differences between the two discursive events' characterization of the disease. A grid of specification is the taxonomic system by which an object is characterized and defined. The grid of specification used for CFS in Bell's speech might be described as a grid of abstract symptoms, a grid in which the disease and its symptoms can be isolated from human carriers for objective analysis. Bell (1990), accordingly, presents the symptoms of CFS in an objective, disembodied fashion. He typically places the symptoms in the subject position of his sentences, which confers upon them a free-standing existence and agency of their own (p. 2). CFS is thus specified primarily as a complex of abstract symptoms, separate from the individual lives of particular human sufferers.

The grid of specification for CFS in the *Newsweek* article, by contrast, is one of personal experience of symptoms. Only one paragraph (Cowley, Hager & Joseph, 1990, p. 62) and one boxed, bulleted list of symptoms (p. 64) treat the CDC disease criteria separately from individual cases. The bulk of the article is replete with a description of the symptoms as they occur in specific patient case histories. The article opens, for instance, with one of these case histories, that of Nancy Kaiser. The case study includes details of Kaiser's personal life: her marital status, hobbies, and medical history. The authors present CFS symptoms not in isolation, as they are presented in Bell's speech, but in the context of Kaiser's experience of them. Kaiser possesses the disease—the disease does not possess her. "She was weak, profoundly tired and plagued by constant bladder infections. Her muscles ached. Her mood shifted unpredictably. Her memory seemed to be failing" (p. 62). When we hear about Kaiser's IQ dropping 22 points, we hear about it in the context of its effect upon the activities in her life. "Her concentration and memory had so deteriorated that she couldn't get through a TV show" (p. 62).

Five other "personal story" vignettes in the *Newsweek* article reiterate the connections between CFS symptoms and human experience. The vignettes are visually emphasized within the article: Each is set off by red bars, includes color photographs of the CFS patients, is typeset differently from the rest of the article, and is flagged with a provocative headline reversed out of a black box: "Policeman's Blues," "Trouble in the Air," "Taunted Teen," "A Dog's Life," and "Spreading the News" (pp. 64, 65, 66, 69, 70). In addition, Cowley, Hager and Joseph draw these vignettes from a variety of geographic locations (New York, Texas, Missouri, California, and North Carolina), indirectly implying the ubiquitousness of human experience of the disease.

The Formation of Subject Positions

The second coordinate of an enunciative field is the formation of enunciative modalities, or subject positions, within a statement or discursive event. The operative question here is not "Who is speaking?" but "What role is being played by the speaker?" Any given individual could step into the same role, since the role is determined not by individual agency and choice but by the conditions of discourse (Foucault, 1972, p. 95). One asks, for instance, what is the institutional site from which the discourse is made? In Bell's case, the site is a formal oral presentation of scientific research results, and the participants at the site are mostly members of the medical community. In the article's case, the site is a written cover story in a national news journal, and the participants members of the general public.

In both of these discursive events, the speaker occupies certain positions with respect to social institutions and forces. In Bell's case, for instance, the speaker's enunciative position is that of a member of the medical research establishment presenting his own research results to other members, and the relationship between him and his listeners is one of community: They are fellow practitioners of medicine. In the article's case, the enunciative position is that of a team of journalists presenting medical research results to the general public, showing how ordinary readers' lives are affected by the results of medical research. The relationship of the journalists to the public is one of community, as people with a mutual interest in public affairs and issues. Both the public and the journalists, however, are outsiders in relation to the medical-scientific community. They are excluded from it, although the journalists do report on the activities of that community.

Because Bell's speech and the *Newsweek* article have different enunciative modalities, the statements they make concerning the place of the research effort with respect to the rest of the medical community differ from one another. In fact, they contradict one another. Bell presents his research as taking place well within the bounds of the medical establishment's research agenda, and he takes care to identify himself closely with the medical-scientific community. Cowley, Hager and Joseph, by contrast, present the same research effort as an adventure story featuring the purported rebellion of the CFS researchers against the medical establishment.

Bell (1990) presents and discusses other research findings on CFS to give an acceptable background for his own work, citing the work of Nancy Climas in Florida on killer cell abnormalities (p. 7), a study linking fibromyalgia and CFS (p. 8), the Peterson-Cheney work with Epstein-Barr virus (p. 10), studies with the HHV6 virus (p. 12), and MRI brain scan research (p. 14). He also traces how his own work is related to clinical research, such as treatment trials for the drug ampligen, in which it was discovered that the 2-4A synthetase pathway in CFS patients is activated on the average of 50 times normal, increasing interleukin-2 levels (p. 15). According to Bell, it was this discovery that led him and his collaborators Cheney and DeFreitas to investigate patients' interleukin-2 levels as a possible key to understanding the etiology of CFS. Thus Bell represents his work as one more research study in a chain of research dedicated to the discovery of scientific truth.

Cowley, Hager and Joseph (1990), on the other hand, depict the work of Bell, Cheney, and DeFreitas as a fairy tale comparable to that of St. George (the researchers) against the dragon (the medical establishment). The story is a quest, a fable, and a mystery, in which the CFS researchers are assigned the role of "good guy" while the medical establishment is assigned the role of "bad guy":

> And a couple of dogged small-town doctors, aided by a Philadelphia virologist, have recently glimpsed what may be the agent behind the affliction. The story of their quest is in one sense a fable about the medical establishment's indifference to problems that defy comfortable assumptions. But it's also a good mystery whose most tantalizing chapter is just now unfolding. (p. 64)

The course of CFS research is presented as a drama of justified rebellion. The story begins as one "dogged small-town doctor," Paul Cheney, is "enjoying a quiet practice" in Incline Village, Nevada, "a prosperous little resort town snuggled against the north shore of Lake Tahoe." The dragon (the medical establishment) begins to appear in the story after Cheney calls the CDC to report an outbreak of an unidentifiable disease. The CDC is slow to respond; two investigators appear only "after much prodding." The upshot of the CDC encounter is a temporary victory for the dragon. "Rather than fretting over Epstein-Barr," the men from CDC suggested, "doctors with sick patients should focus on 'more definable, and possibly treatable, conditions'" (p. 64). Meanwhile, back in Lyndonville, New York, David Bell was dealing with a CFS outbreak of his own and experiencing the same lack of support from the medical establishment. The phrase "trying in vain" (p. 65) indicates the valiant but fruitless attempt of the tenacious doctor to gain support.

Soon a multitude of St. Georges pop into the story. What is described in the *Newsweek* article (pp. 64–66) is not the development of CFS research according to a logical progression of truth within the medical community; rather, it is the cooperative efforts of "a small but devoted cadre of medical researchers" (p. 65), who are positioned against that community.

The Use of Style

A third coordinate of the enunciative field is style. In addressing this coordinate, the archaeologist attempts to discern how concepts are organized within the event as a whole. A brief comparison of the use of graphics in Bell's speech and the Cowley, Hager and Joseph article serves to illustrate how this third aspect of the enunciative field plays itself out in these discursive events. In his speech, Bell presents his data on the effect of CFS on school grades in tabular format, on overhead transparencies, with none of the students named. When he discusses stages of the illness, he does so with lines on graphs indicating functional levels, not with verbal descriptions of human activities. Overall, he presents his immunology data using a quantitative style, which places his data on CFS within the conventional style of science. Numbers, percentages, and statistical tests lend authority to his discussion of cytokines and viruses.

The visuals in the *Newsweek* article create an entirely different impression. Their style is narrative and personal rather than quantitative. Full-color photos of CFS sufferers are the prevalent form of graphic support, supplemented only by a SPECT-scan image and two shaded boxes with bulleted-list descriptions of the illness. In addition, a full-page color photograph of David Bell reinforces the "St. George versus the dragon" implications of the article's text. In this photo, David Bell stands in the foreground, arms crossed, with a group of children standing behind him. We need not stretch the bounds of interpretation very far to see the protector role implied in the composition of the photo.

The Use of Strategy

The final aspect of a statement's enunciative field is strategy. Foucault's use of the word "strategy" must not be confused with definitions typical of the term, which imply authorial intention. To Foucault, a strategy is simply a theme or theory enveloping and reflecting patterns in the formation of objects, subjects, and concepts within a discourse. In some ways, therefore, strategy incorporates aspects of all three of the other fields. Some examples of strategies that Foucault (1972) cites include 19th-century philology's concept of the "kinship of all Indo-European languages," the 18th-century conception of an evolution of the species within the continuity of nature, and the Physiocratic concept of agricultural production as the foundation of all wealth circulation (p. 64).

Foucault's discussion of strategy is the least complete in his presentation of the four aspects of discursive formation.[6] He suggests three potential ways to locate themes or strategies, including establishing points of diffraction (points of incompatibility) within a discursive formation; determining the economy of the discursive constellation; and discovering the function that the discourse plays in relation to nondiscursive practices. The first way applies more to analysis of discourse occurring within one particular discursive formation than to comparative analysis of discourse arising from different discursive formations. The third way requires a thorough analysis of extratextual factors, which is outside the scope of this study. I will therefore restrict my comments here to a discussion of the economy of the discursive constellation as it applies to the strategies manifested in the two events I analyzed.

A discursive constellation is a collection of discursive formations that are contemporary with or related to one another. Within a discursive constellation, a particular discursive formation may be in a number of relationships to other discursive formations: analogy, for instance, or opposition, or complementarity (Foucault, 1972, p. 66). In the case of the CFS speech and the *Newsweek* article, paradoxical relations of opposition and complementarity appear.

At first glance, the discourse strategies appear to be oppositional rather than complementary. Bell's speech, for instance, reinforces the authority of the medical research community to define disease. Common strategies and themes of medical discourse are woven throughout his speech. One of the main medical themes at work is the concept of etiology. What viral or bacterial agent causes the disease? If it can be determined what causes CFS, it becomes an orthodox disease, and Bell presents his study as a move toward determining the etiology of CFS. Another theme or strategy in the speech is that of objectification. The concept of the object, the position of the subject, and the style of the discourse all contribute to a process of objectifying CFS. The disease itself is treated as if it could be isolated from its human sufferers, placed into a laboratory setting, and observed apart from them. As the speaking subject, Bell adopts the position of an inquiring scientist dispassionately performing experiments that fit within a tradition of other experiments. Stylistically, the speech presents information about the disease quantitatively, also implying objectivity. In the *Newsweek* article, by contrast, Cowley, Hager, and Joseph portray Bell in an entirely different light: as a rebellious researcher positioned against the medical establishment. The dominant strategy or theme is one of human subjectivity. CFS is a disease whose symptoms are described as they occur within the context of human experience. Another strategy or theme is that of drama. There is a theatricality to the narrative of the progress of CFS research, a drama that is entirely missing from Bell's speech. The *Newsweek* article is an adventure saga of a small group of doctors (the CFS researchers) who dared to take on the challenge of going against the mainstream medical establishment.

Paradoxically, however, Bell's speech and the *Newsweek* article are complementary as well. Both discursive events create an enhanced level of status and authority for CFS research and researchers within the matrix of institutional power structures. Bell's discursive event creates enhanced status by constructing an identification of CFS researchers with the medical community. Bell's research is described according to acceptable rules of truth for medical discourse and becomes situated within an authoritative tradition of accepted scientific research. The discursive event thus constitutes a shift in status and power within the medical-scientific community by appropriating a more dominant position for CFS researchers, but it at the same time affirms the authority of the medical community itself. Through an entirely different strategy, one that undermines the medical research institution in certain ways by portraying it as both blind to the significance of its own members' findings and insensitive to human suffering, the *Newsweek* article also constitutes the same status shift, enhancing the cultural capital of CFS researchers within the institution of medical research. The article, however, unlike Bell's speech, reduces the overall cultural capital of medicine. In a way, then, since both discursive events increase the perceived value of the work of CFS researchers with respect to the work of other medical researchers, they are complementary nodes in the matrix of power relations within which they emerge; yet they simultaneously formulate competitive versions of medical authority. Bell's speech reinforces the authority of the institution of medicine and asserts its ultimate infallibility, while the *Newsweek* article sends a message that the medical institution can be wrong and implies that it ultimately derives its authority from its service to human beings.

CONCLUSION

The foregoing archaeological critique of Bell's speech and the *Newsweek* article reveal a range of knowledge and power relations. On the surface, the topic of both discursive events analyzed was the same, but the differences between the two discursive events suggest that differing rules governed their formation. Bell's subject position within the medical community required that he reinforce and support the authority of that community to define disease, which he accomplished by using conventional etiological grids of specification to define the disease, placing his CFS research within a progressive series of ongoing medical research studies, and representing his research results quantitatively. Cowley, Hager and Joseph's position as outsiders, by contrast, allowed them to undermine the medical establishment, which they did by creating a subjective formulation of the disease, using graphics that play on human emotion, and valorizing the research team as rebels in an adventure narrative. Both discursive events, however, created an enhanced status position for CFS research and researchers, and in this respect they exist in complementary relation to one another within the discursive constellation.

What are the implications of these findings? First of all, this study supports the contentions of Blair (1987), Cooper (1988), Foss, Foss and Trapp (1991), and Foss and Gill (1987) that Foucault's archaeology can reveal configurations of power and knowledge in society that are not necessarily made visible by traditional rhetorical frameworks of analysis. Using Foucault's theory of the enunciative field and its role in the shaping of discourse, rhetoricians can reorient the focal point of criticism away from an examination of authorial intention within particular discursive events toward larger issues of discourse production and institutional desire, concentrating on which institutional interests are being advanced in discursive events and how those interests are being advanced. This is not to advocate that rhetorical critics abandon other, more traditional frameworks; rather, it is to say that certain questions about discourse may call for different methodological choices.

These findings thus tend to confirm and extend Gaonkar's (1997b) assertion that scientific discourse requires what he calls a "close reading of the third kind," in which a text is examined "not to ascertain the authorial signature, but to map the condition of its existence, the mode of its dissemination, and the forces of its articulation" (p. 352). This critical perspective focuses on how institutional forces and conditions are bound up with what *can* be said. Can a medical professional speaking to other medical professionals in the U.S. do anything other than describe a disease by a taxonomy of abstract symptoms, from a position of alliance with the medical community, through quantitative graphs and tables using strategies of etiology? Must a journalist describe a disease by a taxonomy of human experience, from a position of critical reflection on the medical community, through narratives using strategies of theatricality? Perhaps, and perhaps not; the range of difference among discursive events can itself be a clue to the complex web of power relations that discourse creates and sustains.

It would also appear from this study that science discourse and discourse about science are not discrete discourse worlds but closely interwoven, despite differences in the discursive rules by which they are formed. It is often convenient, for the sake of analysis, to treat the two types of discourse as if they were separate, as I have done in this study. Once the comparisons are accomplished, however, interdependence between the two becomes apparent. As Lyne and Howe (1986) have argued, a two-world model of scientific discourse and public discourse about science cannot adequately explain how discourse is changed, transformed, and deformed in the course of its appropriation by different social groups and communities (p. 142). This study has shown how an instance of science discourse and an instance of discourse about science can both advance the cultural capital of the same social group, which suggests that the two types of discourse can occupy complementary positions within a discursive constellation. Yet it has also shown how the same two instances of discourse can create competing formulations of institutional authority, which suggests that the two types of discourse can occupy competitive positions as well. Since past research has already effectively delineated many of the differences between science discourse and discourse about science, perhaps the challenge for the future is to make their relative positions within the overall architecture of scientific enterprise more evident by plotting the intersections, complementarities, and oppositions of their enunciative fields.

NOTES

[1] For specific responses to and extensions of Gaonkar's article, see *Southern Communication Journal, 58,* entire issue, and Gross and Keith (1997). Campbell and Benson (1996) have also touched on some of Gaonkar's discussion in their book review of major science studies.

[2] A transcript of the speech is available from the author.

[3] Such a task would require analyzing a large number of documents, looking for specific regularities in the discourse.

[4] Foucault does alter this position somewhat at a later point in his career, acknowledging that a statement is similar to a speech act in many ways (Blair, 1987, p. 368).

[5] Kusch (1991) has developed a mathematical equation displaying the relationship of these four factors to the statement (p. 61).

[6] Foucault (1972) attributes this sketchiness to the fact that he has not yet done a major study requiring a detailed analysis of strategies, themes and theories (p. 65).

REFERENCES

Bell, D. (1990, September 12). Chronic Fatigue Syndrome: Illness or illusion? Speech to Capital District Physicians Association and Capital District CFIDS Support Group. Albany, New York.

Blair, C. (1987). The statement: Foundation of Foucault's historical criticism. *Western Journal of Speech Communication, 51,* 364–383.

Blair, C., & Cooper, M. (1987). The humanist turn in Foucault's rhetoric of inquiry. *Quarterly Journal of Speech, 73,* 151–171.

Campbell, J. A. (1986). Scientific revolution and the grammar of culture: The case of Darwin's *Origin. Quarterly Journal of Speech, 72,* 351–376.

Campbell, J. A., & Benson, K. R. (1996). The rhetorical turn in science studies. *Quarterly Journal of Speech, 82,* 74–91.

Cooper, M. (1988). Rhetorical criticism and Foucault's philosophy of discursive events. *Central States Speech Journal , 39,* 1–17.

Cooper, M. (1989). Reconceptualizing ideology according to the relationship between rhetoric and knowledge/power. In C. Kneupper (Ed.), *Rhetoric and ideology: Compositions and criticisms of power* (pp. 30–41). Arlington, TX: Rhetoric Society of America.

Cowley, G., Hager, M., & Joseph, N. (1990, November 12). Chronic Fatigue Syndrome: A modern medical mystery. *Newsweek,* 62–70.

Fahnestock, J. (1986). Accommodating science: The rhetorical life of scientific facts. *Written Communication, 3,* 275–296.

Foss, S. K., Foss, K.A., & Trapp, R. (1991). *Contemporary Perspectives on Rhetoric* (2nd ed.) (pp. 209–239). Prospect Heights, IL: Waveland.

Foss, S. K., & Gill, A. (1987). Michel Foucault's theory of rhetoric as epistemic. *The Western Journal of Speech Communication, 51,* 384–401.

Foucault, M. (1972). *The archaeology of knowledge and the discourse on language.* (A.M.S. Smith, Trans.). New York: Pantheon–Random House.

Foucault, M. (Ed.). (1975). *I, Pierre Riviere, having slaughtered my mother, my sister, and my brother...; A case of parricide in the 19th century*. Lincoln, NE: University of Nebraska Press.

Foucault, M. (1980). The confession of the flesh. In C. Gordon (Ed.), *Power/knowledge: Selected interviews and other writings by Michel Foucault, 1972–1977*. New York: Pantheon.

Foucault, M. (1982). Afterword: The subject and power. In H. L. Dreyfus, & P. Rabinow, *Michel Foucault: Beyond structuralism and hermeneutics*. Chicago: University of Chicago Press.

Gaonkar, D. P. (1993). The idea of rhetoric in the rhetoric of science. *The Southern Communication Journal, 58,* 255–295.

Gaonkar, D. P. (1997a). The idea of rhetoric in the rhetoric of science. In A. G. Gross & W. Keith (Eds.), *Rhetorical hermeneutics: Invention and interpretation in the age of science*. (pp. 25–85). Albany, NY: State University of New York Press.

Gaonkar, D. P. (1997b). Close readings of the third kind: Reply to my critics. In A. G. Gross & W. Keith (Eds.), *Rhetorical hermeneutics: Invention and interpretation in the age of science* (pp. 330–356). Albany, NY: State University of New York Press.

Gross, A. G. (1990). *The rhetoric of science*. Cambridge, MA: Harvard University Press.

Gross, A. G., & Keith, W. (Eds.). (1997). *Rhetorical hermeneutics: Invention and interpretation in the age of science*. Albany, NY: State University of New York Press.

Halloran, S. M. (1984). The birth of molecular biology: An essay in the rhetorical criticism of scientific discourse. *Rhetoric Review, 3,* 70–83.

Killingsworth, M. J., & Palmer, J. S. (1992). *Ecospeak: Rhetoric and environmental politics in America*. Carbondale, IL: Southern Illinois University Press.

Kusch, M. (1991). *Foucault's strata and fields: An investigation into archaeological and genealogical science studies*. Boston: Dordrecht.

Lessl, T. M. (1985). Science and the sacred cosmos: The ideological rhetoric of Carl Sagan. *Quarterly Journal of Speech, 71,* 175–187.

Lyne, J. (1990). Bio-rhetorics: Moralizing the life sciences. In H.W. Simons (Ed.), *The rhetorical turn: Invention and persuasion in the conduct of inquiry* (pp. 35–57). Chicago: University of Chicago Press.

Lyne, J., & Howe, H.F. (1986). "Punctuated equilibria": Rhetorical dynamics of a scientific controversy. *Quarterly Journal of Speech, 72,* 132–147.

Melia, T. (1984). And lo the footprint ... Selected literature in rhetoric and science. *Quarterly Journal of Speech , 70,* 303–313.

Prelli, L. J. (1989). *A rhetoric of science: Inventing scientific discourse*. Columbia, SC: University of South Carolina Press.

part II

A Case Study of a Contemporary Controversy in Science

5

Cold Fusion and the Sociology of Scientific Knowledge*

Trevor Pinch
Cornell University

When two chemists working at the University of Utah announced to the world's press on March 23, 1989, that they had discovered fusion—the controlled power of the hydrogen bomb—in a test-tube, they launched the equivalent of a scientific gold rush. And the gold was to be found everywhere—at least in any well-equipped laboratory. The two scientists were Martin Fleischmann and Stanley Pons.

The apparatus was simple enough (see Figure 5.1): a beaker of heavy water (like ordinary water but with the hydrogen atoms replaced by "heavy hydrogen," otherwise known as deuterium); a palladium "electrode" known as the cathode, and a platinum electrode, known as the anode. A small amount of the "salt," lithium-deuteroxide, was added to the heavy water to serve as a conductor. Though these substances are not in everyday use, and are rather expensive, they are quite familiar to any modern scientist; there is nothing exotic about the apparatus. Put a low voltage across this "cell" for a period up to several hundred hours, and out should come the gold—fusion power. The heavy hydrogen atoms should fuse together into helium, releasing energy; this is the way the sun is powered. The telltale signs of fusion were heat and nuclear byproducts such as neutrons—a sub-atomic particle—and traces of the super-heavy hydrogen atom, tritium.

The experiment seemed straightforward and there were plenty of scientists willing to try it. Many did. It was wonderful to have a simple laboratory fusion experiment

* This chapter is reprinted from Pinch, T. (1994). Cold fusion and the sociology of scientific knowledge. *Technical Communication Quarterly, 3*, 85–100.

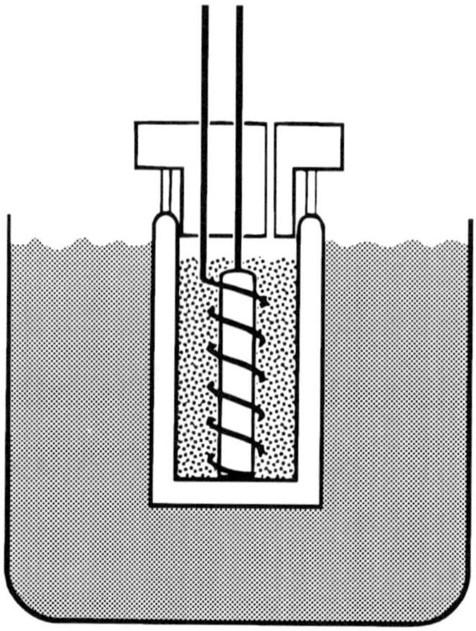

FIGURE 5.1. Cold Fusion Cell. Redrawn by Steven W. Allison from Close (1990, p. 76).

after the decades of embarrassing attempts to control hot fusion. This effort required multi-billion dollar machines whose every success seemed to be capped with an unanticipated failure. "Cold fusion" seemed to provide, as Martin Fleischmann said during the course of that famous Utah press conference, "another route"—the route of little science.

Scientists all over the world immediately started scrambling for information about the experiment. Details were hard to come by. Faxes, electronic mail networks, newspapers, and television all played a role. Some scientists did not wait for details. That same night enterprising students at MIT started the first attempted replications based on a video of a TV news program on which the apparatus had briefly been shown. Such experiments had little chance of success because the exact conditions employed by Pons and Fleischmann were not yet known. Cold fusion experiments were to suffer from their apparent simplicity—at least in the early days before scientists recognized just how complicated a palladium-deuterium electrolytic cell could be.

Within a week a photocopied manuscript showing the technical details of the experiment became available. Now replication started with a vengeance. Scarce supplies of palladium were bought up and pieces of equipment were scavenged from everywhere. Many stayed up all night nursing their electrolytic cells. Science had seen nothing like it; neither had the world's press, which ran continuous news items and updates of progress. It was "science by press conference" as scientists queued up to announce their latest findings and predictions to the media.

And for a while it looked as if cold fusion was going to live up to its promise. Amazingly, in the week following the first announcement it became clear that there was not just one Utah cold fusion group, but two. The second independent group was located at nearby Brigham Young University, and they too had been getting positive results for the previous three years. This group, headed by physicist Steven Jones, had not found excess heat, but had detected neutrons from a cold fusion cell (although at a much lower level than that claimed by Pons and Fleischmann). Both groups had submitted their results to the prestigious scientific journal *Nature.*

Texas A&M University soon announced to the waiting media that they too were seeing excess heat from a cold fusion cell, and then came an announcement from Georgia Tech that they were seeing neutrons. Positive results were reported from Hungary and elsewhere in Eastern Europe. Rumors of positive results came in from all over the scientific world. The Japanese were supposed to be launching their own massive detection program.

Patents had been filed on behalf of Pons and Fleischmann by the University of Utah. Indeed, part of the reason for the press release before the results were published in a scientific paper (a breach of strict scientific etiquette that was to be held against the scientists) was the University of Utah's concern to ensure priority over the nearby group at Brigham Young University. Utah seemed set to become the "gold rush state" with the state legislature meeting and voting that $5 million be allocated for the cold fusion effort. Congress was approached for a possible further $25 million. Even President Bush was being kept apprised of developments.

But then doubts started to surface. It transpired that Georgia Tech had made a mistake; their neutron detector turned out to be heat sensitive. The Texas A&M excess heat measurements were explained away by an improperly earthed temperature-sensitive device. Groups at MIT and national laboratories such as Lawrence Livermore and Oak Ridge were not yet seeing anything. Pons and Fleischmann's paper was mysteriously withdrawn from *Nature.* Congress decided to put the $25 million on hold.

At the American Physical Society meeting that May in Baltimore, with the ever-present media circus in attendance, criticism reached a crescendo. An MIT group claimed that Pons and Fleischmann had incorrectly interpreted their evidence for neutrons; a prestigious California Institute of Technology (Cal Tech) group reported detailed replication attempts, all negative, and cast doubt upon the correctness of the Utah measurements of excess heat; and finally a Cal Tech theorist pronounced that cold fusion was impossible theoretically and accused Pons and Fleischmann of incompetence and possible delusion. The University of Utah pair were not at the meeting to defend themselves, but Steven Jones from the other Utah cold fusion group was there. Unfortunately for Pons and Fleischmann, even Jones distanced himself from their work, claiming that he too had doubts about the measurements of excess heat.

For most of the gathered community of physicists, already skeptical about whether chemists could overturn cherished assumptions about fusion physics, enough was enough. Gold became fools gold, or that at least is how the story goes. As we will see, there is more to be said, much more.

SCIENTIFIC CONTROVERSIES AND THE SOCIOLOGY OF SCIENTIFIC KNOWLEDGE

The cold fusion episode is a dramatic example of a much wider scientific phenomenon—the scientific controversy. Because of the unparalleled public attention the dispute has received and the sheer size of the scientific effort (facilitated by the seeming ease of the experiments), it provides unquestionably the best-documented case to date of a scientific controversy. The study of scientific controversy has played a pivotal role in the development of the field of sociology of scientific knowledge (SSK) (Collins, 1981; Pinch, 1986; Shapin & Schaffer, 1985). It has been argued persuasively that scientific controversies form a strategic research site for studying science. During a controversy, social processes not normally visible within science can become unusually explicit. What counts as a repeatable experiment; the relationship between theory and experiment; the types of scientists who can legitimately contribute to the production of scientific knowledge; what counts as bias, impropriety, and breaches of scientific method; and the role of the media are just a few of the matters which are given concerted attention during a controversy. Under the lens of a scientific controversy, the good, the bad, and the ugly within science come into focus as never before.

In this essay I want to show in a general manner how SSK can be used to illustrate features of the controversy. My approach, although related to the work of rhetoricians on the cold fusion controversy (Gross, 1992; Taylor, 1994), is a much broader attempt to show how cold fusion neatly falls within one well-established and highly influential tradition toward studying science.

Before commencing, it is perhaps wise to outline the guiding assumption of SSK. Although the claims of SSK have been formulated in different ways, and recent years have produced increasing debate in the field (see Pinch, 1991), there is agreement that the key tenet is a methodological prerequisite known as "symmetry" (Bloor, 1991). This requires the sociologist to apply sociological explanation symmetrically to beliefs that are taken to be true as well as beliefs that are taken to be false. Such a stance entails a radical departure from all other earlier sociologies of science because it does not reserve sociological explanation for beliefs which are taken to be false. Social epidemiology and false consciousness are not options within SSK. Beliefs that are taken to be true are shaped or constructed by the same social processes that shape false beliefs.

The application of symmetry has dramatic effects. It means, for instance, in the cold fusion controversy that we cannot separate the wheat of genuine science from the chaff of error. Both sides in the dispute have to be treated equally. That is to say we cannot prejudge which side is right. An SSK-inspired study of cold fusion would aim to try and recover in as much detail as possible the rationales, rhetorics, and argumentative stances of the cold fusion proponents and their critics as the controversy unfolds.

The application of aspects of SSK to cold fusion has already begun in the scholarly literature (Gieryn, 1992; Lewenstein, 1992). My story here focuses attention on how the two sides in the dispute are able to maintain their respective positions despite their concerted efforts to bring about a resolution. In other words, we will examine what has become known as the "interpretative flexibility" (Collins, 1981; Pinch, 1986) of scientific claims. We will look at how the "experimenters' regress" (Collins, 1992) prevents an experimental resolution of the controversy by attempts at replication. We will touch upon the open-ended nature of the theoretical dispute as to whether cold fusion is or is not theoretically possible. We will also draw attention to features that seem to be leading to a closure of the controversy.

THE LITTLE SCIENCE ROUTE TO FUSION

It is palladium, or rather one property of palladium, that provided the impetus to the search for cold fusion. It is known that palladium has a surprising ability to absorb vast quantities of hydrogen. If a piece of palladium is "charged" with as much hydrogen as it can absorb, then the pressure inside the crystal lattice dramatically increases. Perhaps at such high pressures the normal barrier of positive charge (known as the Coulomb barrier) preventing nuclei coming together to fuse could be overcome. It was a long-shot, but scientists in the 1920s had actually tried to produce fusion between hydrogen nuclei this way.

Pons and Fleischmann were unaware of the earlier work when they started their experiments in 1984. Martin Fleischmann is one of Britain's most distinguished electrochemists. Stanley Pons (an American) visited the University of Southampton to study for his PhD in 1975, and that is where the two met. Fleischmann, who was Faraday Professor of Electrochemistry at Southampton, had a reputation for being the sort of scientist who liked to carry out high-risk science, pursuing bold and innovative ideas and approaches. Indeed, he has built his career with such work and some of the risks have paid off. Fleischmann has made a number of important discoveries, as recognized by his election as a Fellow of the Royal Society in 1986.

The reason Fleischmann was in Utah in 1989 had to do with British Prime Minister Margaret Thatcher. The Thatcher cutbacks of funding to British Universities in 1983 meant that Fleischmann was forced to take early retirement from the University of Southampton (where he retained an unpaid position). He became a freelance researcher teaming up with Pons, who was by now a productive scientist in his own right and chair of the University of Utah chemistry department. Pons was 46 and Fleischmann 62 when the discovery was announced.

JONES' INVOLVEMENT

The March 23 announcement cannot be understood without reference to the work of the other Utah group, led by Steven Jones at Brigham Young University. Although the scientific community was unfamiliar with Pons and Fleischmann's work on cold fusion, they had been following Jones' progress on particle-induced fusion for several years. Like hot fusion research, particle-induced fusion was a frustrating step away from the excess energy output needed for commercial exploitation.

In 1985 Jones began cold fusion research using an electrolytic cell. He built a low-level neutron detector to measure any fusion that was occurring. In 1986 he started to observe neutrons at a rate just above background levels. By 1988, using an improved detector, he felt confident that he had found definite evidence of neutron production.

Jones had carried out this research unaware of the similar efforts being carried out at the nearby University of Utah. He first heard of Pons and Fleischmann's experiments in September 1988 when the Department of Energy sent him their research proposal to referee.

It was unfortunate for both groups that such similar work was being pursued in such close proximity. In view of the obvious commercial payoff that might result from cold fusion and the need for patent protection it meant that a certain amount of rivalry and suspicion arose between the two groups. Exactly what they agreed regarding the joint publication of their results is still disputed.

THE CONTROVERSY

Most of the cold fusion controversy has concerned the results of Pons and Fleischmann. The levels of neutrons detected by Jones were orders of magnitude less, and he has never claimed to observe excess heat. Jones' results also did not pose the same theoretical challenge. Furthermore, Jones, unlike Pons and Fleischmann, has made a point of playing down any commercial application.

The detrimental effect on the credibility of scientists' findings, when they get caught up in a scientific controversy not of their own choosing, is revealed by the reception of Jones' results. Few would doubt, given his previously established reputation in the field, the minimal theoretical consequences his results posed, and the modest manner in which he presented them, that if it had not been for Pons and Fleischmann, Steve Jones would by now have quietly established an interesting fact about the natural world—fusion of small amounts of deuterium in palladium metal.

Despite his attempts to distance himself from the other Utah group, Jones has inevitably been subject to criticism. The reality of his neutron measurements has come into question, and there is no consensus as to whether he has observed fusion.

Pons and Fleischmann, unlike Jones, had no established reputation in the field of fusion research; they were chemists, not physicists. Moreover they were claiming something that to most physicists was theoretically impossible. Not only did it seem extremely unlikely that fusion could be occurring but, also, if all the heat excess was caused by fusion then the levels of neutrons produced should have been more than enough to kill Pons and Fleischmann and anyone else who happened to be in the close proximity of one of their cells. In short, fusion could not occur, and if it did, then they should be dead. This is what one might refer to as a "knockdown" argument!

Nevertheless, Pons and Fleischmann had two sorts of evidence to back up their claims: excess heat and nuclear products.

Excess Heat

Testing for excess heat is in essence no more than a high school physics problem. A careful account of the power input and output of the cell is kept, including all the known chemical reactions that are capable of transforming chemical energy into heat. This accounting needs to be carried out over a period of time because at any one moment the books might not balance as energy might get stored in the cell (turning it into a heat bank, as one might say). It is a fairly straightforward procedure to establish the power output by measuring the temperature rise, the cell first having been calibrated using a heater of known power. In practice the experiment took some time to perform because the palladium electrodes had to be fully charged with deuterium.

The heat excess varied between cells. Despite the capricious nature of the phenomenon, Pons and Fleischmann were confident that the heat excess could not be explained by any known chemical process or reaction.

Nuclear Products

The most direct proof of fusion would be the production of neutrons correlated with the excess heat. The first neutron measurements attempted by Pons and Fleischmann were relatively crude. The output from a cell was compared with the background as measured at a distance of 50 meters from the cell. A signal three times the background was recorded for this one cell. This was a suggestive result, but as neither the energy of the neutrons was known, nor whether the background was the same close to the cell as at 50 meters distance, it was far from conclusive. A more satisfactory procedure was to measure the gamma ray bursts produced by neutrons captured by protons in the water bath surrounding the cell. Unfortunately, the numbers of neutrons detected (if any) were billions less than would be expected if all the heat was produced by the deuterium fusion reaction.

Another piece of evidence for fusion having taken place would be the presence of its products, such as tritium. Pons and Fleischmann found traces of tritium in the palladium cathode of one cell. The difficulty with this finding—a problem that has beset all the claims—is that tritium is a known contaminant of heavy water.

REPLICATION

As mentioned already, after the March 23 announcement, attempts to repeat the experiments followed fast and furious. Although they got much media attention, these early results (both positive and negative) counted for little. The embarrassment caused by the premature announcements from Georgia Tech and Texas A&M cautioned those scientists who were seriously trying to repeat the experiment that they faced a long struggle.

Part of the difficulty facing scientists trying to repeat the experiment was that Pons and Fleischmann's account of what they had done was insufficiently detailed. The exact size of the electrodes to be used, the current densities at which to operate the cells, whether the lithium salt was crucial or could be substituted by another salt, whether the cathode was "poisoned" and with what, and for how long the experiment should be run were all unclear. Following on from their initial announcement, Pons and Fleischmann were inundated with requests for information. In the frenetic atmosphere at Utah, scientists did not always find it easy to get the crucial information.

Some have accused Pons and Fleischmann of deliberate secrecy in order to secure patent rights or (later on, when many became disillusioned) to hide their own incompetence. However, given the commercial importance of the discovery, securing the patent rights is no small matter; it is routine in areas of biotechnology. Also, it seems that Pons and Fleischmann were initially hesitant because of their own uncertainties and their fears about the dangers of the experiment.

The elusive details of the experiments spread through an informal network of electronic mail and telephone contacts. Indeed, electronic mail may have been important in this controversy in producing the rapid consensus against cold fusion that developed after the Baltimore American Physical Society meeting. For instance, Douglas Morrison, a CERN physicist and early enthusiast of cold fusion, set up an electronic newsletter that seems to have been widely read. Morrison soon became skeptical of the claims and referred scientists to Irving Langmuir's notorious talk on "pathological science," where a number of cases of controversial phenomena (including N-Rays and ESP) in science were dismissed as a product of mass delusion. (Langmuir's talk was reproduced in *Physics Today* in October 1989.) Cold fusion was, according to Morrison, the most recent case of pathological science.

What became clear early on was that although most groups saw nothing, a few had positive results. The classic problem of replication during a scientific controversy was surfacing. Negative results could be explained away by the believers due to crucial differences between the original experiments and the copies. To those who failed to find anything, however, this was simply confirmation that there was nothing to be found. Fleischmann and Pons's own attitude, as expressed in their testimony to Congress in April 1989, was that they were not surprised by the negative results as many cells were being set up with incorrect parameters and dimensions.

The problem of replication is part of a circular pattern of reasoning which is always manifest during scientific controversies. Harry Collins (1992) has termed this the "experimenter's regress." The problem is that doing an experiment is an exercise in skill and it is not clear during a controversy how the necessary skill is to be judged. For the believers, a competently performed experiment is one that will produce the phenomenon. For the critics a competently performed experiment is one that is negative. Because whether cold fusion exists or not depends on which experiments are judged competent—and there is no independent measure of competence—experimenters are thrown into a regress. The importance of this is that it shows that experiments alone are never capable of settling a controversy; other factors always have to be brought in to try to break the regress.

Of the early positive replications reported, one of the most important came from Robert Huggins, a materials scientist at Stanford University. Huggins had run two cells, one with ordinary water and one with heavy water, and found that only the heavy water cell produced excess heat. This answered a long-running criticism of Pons and Fleischmann for not setting up a "control" cell with ordinary water. Huggins has consistently found positive results over the years. However, for the critics, such results are not definitive because other experimenters find excess heat from ordinary water cells, indicating that the effect cannot be from nuclear fusion of deuterium.

Of the negative results, one of the most influential has come from a Cal Tech group headed by chemist Nathan Lewis and physicist Charlie Barnes. The Cal Tech team tried a variety of combinations of conditions and found nothing. As mentioned already, Lewis reported the negative results to the Baltimore American Physical Society meeting to dramatic effect. His results had extra impact because he implied that Pons and Fleischmann were guilty of an elementary oversight. They had neglected to stir the electrolyte thus allowing hot spots to develop and produce spurious temperature readings.

However, it seems that Lewis's charges were misplaced. Pons and Fleischmann claimed that there was no need to stir the electrolyte because the deuterium bubbles produced by the reaction did the job sufficiently well. In order to demonstrate the error, Lewis had tried to make an exact copy of Pons and Fleischmann's cell. He had taken as his source a photograph of a cell in the *Los Angeles Times.* It turns out that this cell was only used by Pons and Fleischmann for demonstration purposes and was of much larger dimensions than the cells used in actual experimental runs. Pons and Fleischmann were later able to demonstrate with a simple experiment—placing a few drops of dye in the electrolyte—that the bubbles acted as an adequate stirring mechanism. But as so often in scientific controversies, Pons and Fleischmann's position on stirring has in turn been challenged by Lewis.

What is taken by most people to be a "knock-down" result turns out, on closer examination, to be subject to the same kinds of ambiguities as the results it claims to demolish. If Lewis's measurements had been unpacked in the same kind of detail reserved for Pons and Fleischmann they might not have seemed as compelling as they did at the time. In the atmosphere of the Baltimore meeting, where physicists were baying for the blood of the two chemists, and where a whole series of pieces of negative evidence was presented (see below), Lewis was able to deliver the knock-out blow. The rhetorical presentation of Lewis' results was particularly important. As I have argued elsewhere (Pinch, 1995), Lewis and his Cal Tech colleague Steve Koonin both used a highly effective form of debunking rhetoric in their talks to ridicule Pons and Fleischmann.

The classic problem of replication has surfaced with another set of highly influential negative findings, those reported by Harwell. As a result of Fleischmann's contact with Harwell, David Williams, a former graduate student of Fleischmann's, actually started his experiments before the March announcement. His results, for all intents and purposes, killed off cold fusion in Britain. Again, on the face of it, the experiments look impressive with a number of cells being checked for excess heat and neutrons over lengthy periods.

The results, however, are not compelling for proponents of cold fusion, such as Eugene Mallove (1991), who claim that almost half the cells were run at currents below the threshold for cell activity. Other criticisms have been made of the Harwell methods of heat estimation. Despite the differing interpretations of the Harwell experiment, for many scientists it was the last word on cold fusion.

The difficulty that the proponents face in getting positive results accepted is well illustrated by the fate of the tritium measurements. It will be recalled that Pons and Fleischmann had themselves found traces of tritium. More evidence came from other experimenters, including a group in India with a long history of making tritium measurements, a group at Los Alamos, and a third group at Texas A&M University. However, since tritium is a known contaminant of heavy water, there is a ready-made "normal" explanation for all such results. It has proved to be impossible to satisfy the critics that no contamination has occurred, because they can always think of ways in which tritium might get into the cell.

It has even been suggested that the affair has involved fraud. In 1990, an article in the journal *Science* puts forward fraud as a factor in the Texas A&M tritium measurements. The impasse between proponents and critics, a situation made worse by each side accusing the other of "unscientific" behavior, is typical of scientific controversies. The critics cite a preponderance of negative results as grounds to dismiss the controversial phenomenon and any residual positive results are explained away as incompetence, delusion, or even fraud. The proponents, on the other hand, account for the negative results due to the failure to reproduce exactly the same conditions as used to obtain positive results. Experiments alone do not seem capable of settling the issue.

COLD FUSION—A THEORETICAL IMPOSSIBILITY?

Most of the debate has been fought out in a setting in which cold fusion has been held to be impossible on theoretical grounds. Cal Tech theorist Steve Koonin, in reviewing all the theoretical possibilities at the May 1989 American Physical Society meeting, was able to make theoretical justifications seem preposterous. There is no doubt that Koonin represents the standard view, but as is typical for a scientific controversy where an experiment seems to go against prevailing theory there is more to say than this. Indeed throughout the cold fusion episode a number of suggestions have been made as to how fusion might occur on the necessary scale and, furthermore, how it might occur without neutrons being produced. Some of the more serious suggestions have come from physics Nobel Laureate Julian Schwinger and the MIT laser physicist who helped invent the X-ray Laser, Peter Hagelstein. One idea has been to think of ways whereby a rare neutronless fusion reaction could be the source of the excess heat, with energy being transferred into the palladium lattice. Hagelstein, drawing upon ideas in laser physics, has also proposed "coherent fusion," whereby chains of fusion reactions occur in a kind of domino effect. However, theory, like experiment, does not provide a neutral Archimedean point whereby the dispute may be resolved.

With the experimental results under a cloud, however, most theorists saw little reason to entertain such exotic ideas. There is no doubt that a theorist such as Hagelstein himself takes these alternative theories very seriously—he has even applied for patents for devices based upon his theories. The risks in pursuing such theories beyond the mere "what if" stage (i.e., treating such theories as serious candidates rather than mere speculations) are illustrated by the Hagelstein case; there have been persistent rumors that his tenure at MIT was in jeopardy after he started to produce theoretical explanations of cold fusion.

CREDIBILITY

The struggle between proponents and critics in a scientific controversy is always a struggle for credibility. When scientists make claims that are literally "incredible," as in the cold fusion case, they face an uphill struggle. The problem Pons and Fleischmann had to overcome was that they had credibility as electrochemists but not as nuclear physicists. And it was in nuclear physics where their work was likely to have its main impact.

Any claim to observe fusion (especially made in such an immodest and public manner) was bound to tread upon the toes of the nuclear physicists and fusion physicists who had already laid claim to the area. A vast amount of money, expertise, and equipment had already been invested in hot fusion programs, and it would be naive to think that this did not affect in some way the reception accorded Pons and Fleischmann.

This is not to say that the fusion physicists simply rejected the claims out of hand (although a few did), or that it was merely a matter of wanting to maintain billion dollar investments (although with the DOE threatening to transfer hot fusion funding to cold fusion research there was a direct threat to their interests), or that this was a matter of the blind prejudice of physicists over chemists (although some individuals may have been so prejudiced), it was simply that no scientist could hope to challenge such a powerfully established group without having his or her own credibility put on the line. As might be expected, the challenge to Pons and Fleischmann has been most acute in the area where the physicists felt most at home, the area of the neutron measurements.

NEUTRON MEASUREMENTS

For many physicists it was the neutron measurements that provided the best evidence for fusion. Yet paradoxically these measurements formed the weakest link in Pons and Fleischmann's claims. The measurements were carried out belatedly and under pressure. Worse, neither Pons nor Fleischmann had any special expertise in such measurements.

It was at Harwell, at a seminar given by Fleischmann shortly after the March announcement, that the first inkling of difficulties was to arise. Fleischmann presented the evidence of neutrons and showed a graph of the gamma ray peak from the water shield. To physicists in the audience familiar with such spectra, the peak appeared to be at the wrong energy. The peak was at 2.5 MeV whereas the expected peak for gamma rays produced by neutrons from deuterium should have been at 2.2 MeV. It looked as if something had gone awry with the calibration of the gamma-ray detector, but it was impossible to tell for certain because Fleischmann did not have the raw data with him and had not made the measurements himself. In any event, by the time the graph appeared in Pons and Fleischmann's first publication, the peak was given at the correct value of 2.2 MeV.

Whether the two versions arose from "fudging" or genuine errors and doubt over what had been measured is unclear. Frank Close, in his much publicized skeptical book about the cold fusion controversy, *Too Hot to Handle* (1990), suggests that the graph was deliberately doctored—a charge taken up by science journalist William Broad (1991) in an article in the *New York Times.* Such accusations should, however, be treated with caution. Close, in particular, falls into the trap of recovering all the gory detail of the proponents' experiments, leaving the critics' experiments to appear as clear-cut and decisive. Such a one-sided narrative merely serves to reaffirm the critics' victory.

The neutron measurements soon came under further scrutiny. Richard Petrasso, of the MIT Plasma Fusion Center, noticed that the shape of the gamma ray peak looked wrong. Preliminary reports of the work were given by Petrasso at the Baltimore meeting to maximum rhetorical effect. In tandem with the Cal Tech negative results, Petrasso's work was to have the decisive impact on the course of the controversy that we have already charted.

The criticism of the neutron measurements was eventually published in *Nature* along with a reply from Pons and Fleischmann. The Utah pair were now placed on the defensive. They published their full spectrum showing no peak at 2.2 MeV but claiming evidence for a new peak at 2.496 MeV. Although they could not explain this peak in terms of a known deuterium fusion process, they maintained the peak was produced by radiation from the cell. Many scientists have taken this episode as showing that the main argument in favor of fusion had collapsed. However, another interpretation is possible. This is that the best evidence for cold fusion always came from the excess heat measurements—the experimenters' own strength. The hastily performed nuclear measurements have always been puzzling because too few neutrons were observed. In trying to "come clean" on the difficulties of interpreting their nuclear measurements, Pons and Fleischmann were attempting to draw attention back to the main thrust of their argument—the excess heat measurements. Indeed, when Pons and Fleischmann published their full results finally in July 1990, the paper was almost entirely about calorimetry—no nuclear measurements were reported.

The trouble was that for many physicists the nuclear data were what had got them excited in the first place and the weakness of the neutron evidence left the excess heat measurements as mere anomalies, possibly of chemical origin. Furthermore, the problems over the nuclear measurements could easily be taken to demonstrate Pons and Fleischmann's incompetence as experimenters. Despite the Utah pair being widely acknowledged as experts in electrochemistry, this kind of "guilt by association" seems to have paid off for the critics and has helped discredit the experiment as a whole.

CONCLUSION

In this account I have focused mainly on the early stages of the controversy, pointing in particular to the role of the Baltimore American Physical Society meeting where the tide turned against Pons and Fleischmann. Today scientists still continue to work on the phenomenon; some positive results are reported, and conferences are held on such topics as "Anomalous Phenomena in the Palladium–Deuterium Lattice." Indeed, the very labeling of the phenomena in this way reflects a feature familiar from other controversies where, in order to try to get the controversial phenomenon accepted, proponents play down its implications for other scientists (Pinch, 1985). Gone are the bold claims that the phenomenon is definitely fusion, and gone is the prospect of a new source of commercial energy just around the corner (although Japanese companies, such as Toyota, continue to support Pons and Fleischmann and invest heavily in cold fusion research). It may eventually be established that there is something unusual going on in the palladium–deuterium lattice but that something is unlikely to be cold fusion as it appeared circa March 1989. The failure in 1989 to attract significant Department of Energy funding means that compared with its initial promise cold fusion research is in decline—at least in the United States. The Utah National Cold Fusion Institute was dissolved in June 1991.

Despite this, and as is true for most scientific controversies, a small group of active proponents continue to "fight the good fight." For the believers the consensus developing against them is evidence of a conspiracy of blind and arrogant scientists aligned with the main funding agencies and organs of scientific publication such as *Nature* magazine. For the critics the fact that some scientists continue to believe in cold fusion is evidence that such people must be motivated by extra-scientific considerations or have invested too much already in this blind alley to be able to back out gracefully.

Both sides claim the facts are with them, both sides impute illicit extra-scientific reasons to the other. There could scarcely be a better case for the relativistic sociology of scientific knowledge.

In the cold fusion controversy the stakes were high. The normally hidden process of "science in action" (Latour, 1986) has become visible. The cold fusion episode is often taken to show that there is something wrong with the way science operates. It is said that scientists claimed too much, based on too little, and in front of too many people. Press review replaced peer review. False hopes of a new age of limitless energy were raised, only to be dashed.

Such an interpretation is unfortunate. Pons and Fleischmann appear to have been no more greedy or publicity-seeking than any prudent scientists who think they have a major discovery with a massive commercial payoff on their hands. The securing of patents and the fanfare of press conferences are inescapable parts of modern science, where institutional recognition and funding are increasingly important. But science itself has not suffered. It is our image of science that needs changing, not the way science is conducted.

REFERENCES

Bloor, D. (1991). *Knowledge and social imagery* (2nd ed.). Chicago: University of Chicago Press.

Broad, W. J. (1991, May 17). Cold-fusion claim is faulted on ethics as well as science. *New York Times*, p. 1.

Close, F. (1990). *Too hot to handle: The race for cold fusion*. London: Allen.

Collins, H. M. (Ed.). (1981). Knowledge and controversy. *Social Studies of Science, 11* (1).

Collins, H. M. (1992). *Changing order* (2nd ed.). Chicago: University of Chicago Press.

Gieryn, T. (1992). The ballad of Pons and Fleischmann: Experiment and narrative in the (un)making of cold fusion. In E. McMullin (Ed.), *The social dimensions of science* (pp. 217–243). Notre Dame, IN: Notre Dame University Press.

Gross, A. (1992). Building neo-Aristotelian theory: The cold-fusion controversy as a test case. Unpublished manuscript.

Latour, B. (1986). *Science in action*. Cambridge, MA: Harvard University Press.

Lewenstein, B. (1992). Cold fusion and hot history. *Osiris 7,* 135–163.

Mallove, E. (1991). *Fire from ice*. New York: John Wiley.

Pinch, T. J. (1985). Towards an analysis of scientific observation: The externality and evidential context of observational statements in physics. *Social Studies of Science 15,* 3–36.

Pinch, T. J. (1986). *Confronting nature: The sociology of solar-neutrino detection*. Dordrecht, Holland: Kluwer.

Pinch, T. J. (1991). The role of scientific communities in the development of science. *Impact of Science on Society 159,* 219–225.

Pinch, T. J. (1991). Rhetoric and the cold fusion controversy: From the chemists' Woodstock to the physicists' Altamont. In H. Krips, J. E. McGuire, & T. Melia (Eds.), *Science, reason, and rhetoric* (pp. 153–176). Pittsburgh and Konshanz: University of Pittsburgh Press and Konshanz University Press.

Shapin, S., & Schaffer, S. (1985). *Leviathan and the air-pump: Hobbes, Boyle, and the experimental life*. Princeton, NJ: Princeton University Press.

Taylor, C. A. (1994). Science as cultural practice: A rhetorical perspective. *Technical Communication Quarterly 3,* 63–77.

6

Science as Cultural Practice: A Rhetorical Perspective*

Charles Alan Taylor
Drury College

Theoretical attempts to specify those criteria that demarcate science from other intellectual activities frequently have been advanced and, just as frequently, found wanting (Gieryn, 1983; Holmquest, 1990). Similarly, among the most striking characteristics of recent rhetorical studies has been the attempt to come to grips with scientists and to state just what it is that makes them rhetorical. Now that the initial critical euphoria surrounding the rhetoric of science has abated a bit, one senses a belief that we have reached a crucial point, one at which we will either continue to spin our theoretical wheels or at which we as a community will develop more coherent programs of research. In a recent review essay, Carolyn Miller (1989) observed that even the best work written under the rhetoric of the human sciences rubric has come "without much advance in theoretical terms" (p. 103). Participants in the rhetoric of science seminar at the 1991 meeting of the Speech Communication Association expressed misgivings about our ability to transcend strict disciplinary concerns so as to speak to the interdisciplinary science-studies community. Indeed, the call for submissions for this volume of TCQ contends that the rhetoric of science literature is mired in "theoretical disarray" (Speech Communication Association, 1991, p. 7).

In one sense, this might be read as a salutary development, signaling a potentially healthy reflexive moment wherein we can rethink (and I hope reshape) our critical

* This chapter is reprinted from Taylor, C. A. (1994). Science as cultural practice: A rhetorical perspective. *Technical Communication Quarterly, 3*, 67–81.

practices. However, it bodes ill if it becomes either an excuse to abandon the search for fruitful theoretical orientations because of the lack of an underlying theme or if it leads to rhetorical foundationalism as a way to reclaim the disciplinary recognition that has too long been denied rhetoric.

In this chapter, I want to illustrate what I take to be a fruitful heuristic for the rhetoric of science literature. In admittedly broad strokes, I want to sketch a conception of science as practice and attenuate extreme versions of the "science is rhetorical" shibboleth; at the same time, I want to broaden the range of discourses relevant for analysis by critics of scientific rhetoric. A brief case study of two dimensions of the 1989 "cold fusion" controversy illustrates how these commitments are instantiated in critical practice. A final section will consider the potential objections to and theoretical implications of this perspective with an eye toward broader theoretical lessons for the rhetoric of science literature.

CRITICAL PERSPECTIVES ON SCIENCE

To date, most rhetorical studies of science seem to arise from viewing their object of investigation (science) from one of two major perspectives. On the one hand, several studies have illustrated convincingly the hidden rhetorics inherent in knowledge claims from Darwin's *Origin* (Campbell, 1986) to Wilson's *Sociobiology* (Lyne, 1990; Lyne & Howe, 1990). In these, science is portrayed as something of a "black box," from whose mysterious interior issue such claims as "e=mc2" or "genes explain much of human behavior." These knowledge claims then become the objects of rhetorical analysis.

The key concern for the second dominant perspective for viewing science seems to be the ways in which various sets of claims are authorized as epistemically secure. Consider Lawrence Prelli's (1989) extensive discussion of the topoi of science. Although he obviously is concerned with sets of scientific claims, his primary critical focus is on how those claims are (de)legitimized. He maintains, for example, that "to be judged reasonable and persuasive in any specific situation, scientific discourse must be perceived as identifying, modifying, or solving problems that bear on a specific scientific community's maintenance and expansion of their comprehension of the natural order" (p. 122). In this case, the deliberative activities of the "audience" of scientific practitioners represent the method by which various claims are warranted (Overington, 1977).

These perspectives have yielded much valuable theorizing and criticism to which my analysis is beholden. Here, however, I want to move off in somewhat different theoretical and critical directions. Although I am under no delusions that the alternative perspective developed below can solve the myriad difficulties confronting rhetorical studies of science, I suggest that a view of science as a series of cultural practices has the potential simultaneously to broaden and to focus our future studies. Here I will borrow liberally from certain constructivists in the sociology of scientific knowledge, albeit with a decidedly rhetorical twist.

Two key concepts merit definition. First, what is meant by culture? Andrew Pickering (1992) characterizes "culture" as "the field of resources that scientists draw upon in their work" (p. 3, fn1). He suggests that a meaningful conception of science can—indeed must—extend beyond the walls of the laboratory. The resources on which scientific practices are dependent often originate elsewhere, whether in grant application deliberations, budgeting debates, or coffee-klatsch discussions.

The second key term here is "practice." Like "discourse," practice has become something of a buzzword in contemporary intellectual circles. Here, I intend only its simplest interpretation, as the act of doing something. In short, then, scientific practice can be understood as what scientists (and those who purport to speak on their behalf) do as they attempt to navigate the many exigencies that impinge on their professional activities.

What is missing from these sociological accounts is crucial, and is precisely the defining characteristic of rhetorical studies. Not surprisingly, what constructivist accounts of scientific practice lack is a rhetorical consciousness, an awareness of the transformative capacity of human discourse. The illumination of discursive processes as central to conceptualizing, enabling, and justifying ongoing practical activities, long thought the purview of the SSK, is the underlying commitment of this perspective on the rhetorical analysis of scientific discourse.

A primary contribution of this orientation is that it simultaneously expands the range of what we can reasonably consider to be rhetorics of science. For example, Congressional budget hearings to revive the Superconducting Super Collider project, in addition to their generic status as political debates, can be understood as scientists' rhetorical practices intended to secure particular resources without which lab practice would be impossible. Such a view encourages us to move beyond the traditional exemplary texts of science, for example, journal articles or "revolutionary" documents such as Darwin's *Origin.* By casting our analytical net more widely to capture discourses that we have generally considered peripheral to "real" science, we can begin to come to grips with science as a complex network of cultural practices rather than simply laboratory practices and the claims issuing from them.

If scientific culture is rhetorical, we might be tempted to assume that scientific practice must reflect what Alan Gross (1991) called a "rhetoric of science without constraints" (p. 283) and what J. E. McGuire and Trevor Melia (1991) described as a "radical rhetoric of science" (p. 301), that is, one that holds that all of scientific practice is rhetorical, without remainder. I think we must resist this temptation. What I detailed above is intended as a perspective for viewing science, not necessarily for critical studies of scientific discourse. Although a commitment to the perspective certainly expands the range of discourses considered scientific, it does not entail that all of the practices of science are rhetorical. Technical practices such as constructing bubble chambers or preparing brain extracts do not strike me as rhetorical, or even as remotely discursive. Certainly the network of propositions which alternately motivate and issue from those behaviors is rhetorical, but the acts themselves appear to be both nontrivial and nonrhetorical. Or, perhaps more precisely, they would be studied more productively from perspectives other than those adopted by most rhetorical critics.

Adopting a view of science as cultural practice, then, simultaneously broadens the range of rhetoric that "counts as" scientific, while maintaining a focus on discursive phenomena. In the next section, I will briefly illustrate how such an orientation might be worked out in critical practice. A case study of the early stages of the 1989 cold fusion controversy will suggest that traditional views of science as restricted to laboratories and peer-reviewed journals cannot account adequately for the myriad rhetorical exigencies confronting contemporary scientific rhetors.

THE SCIENTIFIC RHETORICS OF COLD FUSION

The story is by now well known. In the wake of the Exxon Valdez disaster and its siren call for a responsible and safe energy policy, University of Utah chemists B. Stanley Pons and Martin Fleischmann announced at a March 23, 1989, press conference that they had conducted experiments that had resulted in "sustained n[uclear] fusion for the first time." The world was told that the "breakthrough process has the potential to provide an inexhaustible source of energy" (Bishop & Wells, 1989, p. 1).

Beyond its ecological potential, this was stunning news for the scientific community. Rather than the multimillion dollar high-temperature magnetic plasma containment devices (tokamaks) or processes of laser bombardment long thought necessary to achieve fusion, the Utah experiment was said to require only a laboratory flask filled with heavy water, lithium salts, a palladium rod encircled by a platinum rod, and an electrical source. Pons and Fleischmann claimed to have achieved heat outputs of up to four times the amount of energy imported into the system (Fleischmann & Pons, 1989; Fleischmann, Pons, & Hawkins, 1989; Fleischmann, Pons, & Hoffman, 1989).

Cold fusion probably reached its halcyon peak with the appearance of a Utah contingent before the House Sub-Committee on Science, Space, and Technology (U. S. House of Representatives, 1989). Ultimately, however, the furor over cold fusion receded in the face of mounting negative evidence (Rothman, 1990). The November 26, 1989, release of the Energy Research Advisory Board's Final Report seemed to deal the death blow to the Utah program by generally dismissing it as anomalous and recommending that it receive no special federal funding.

I want here to focus on the all-important first eight months of the controversy when the scientific community seemed to make confirmation or repudiation of the Utah reports a top priority (Close, 1991; Greenberg, 1989; Mallove, 1991; Peat, 1989). Modifying and extending previous analyses (Hynes, 1989; McAllister, 1992; Taylor, 1991a; Taylor, 1991b), I consider the claims advanced by critics and supporters of the Utah experiments regarding whether Pons and Fleischmann had comported themselves in accordance with the cultural commitment to "intellectual communism" (Merton, 1973, pp. 269–278). Although I do not share Robert Merton's functionalist motives (Collins & Restivo, 1983, pp. 187–188), analysis of this set of rhetorical practices reveals both the flexibility of this ostensibly field-invariant cultural vocabulary and its seeming inability to account adequately for the important disciplinary conflict characteristic of this episode.

Much criticism was directed at what was perceived by many as the Utah researchers' violation of the normative commitment to appropriately communal behavior. Two specific manifestations of this rhetorical practice were the indictment of their insufficient disclosure of relevant technical details and of their failure to abide by the norm's structural "guarantor": peer review procedures.

Several critics charged that Pons and Fleischmann had failed (or chosen not) to provide adequate details in their initial *Journal of Electroanalytical Chemistry and Interfacial Electrochemistry* publication (Fleischmann & Pons, 1989) as well as in subsequent appearances before Congress and several professional conferences. Oakland University physicist Craig Sevilla insisted that there were only two possibilities: "Either it's a bunch of hokum … or they are holding out a critical piece of information" (as cited in McCann, 1989, p. F11). His colleague, Craig Taylor, offered a third possibility while still condemning the dearth of experimental detail, saying that "instead of deliberately taking something out of the recipe, it might have been an oversight on their part because of the tremendous pressure they were under" (as cited in McCann, 1989).

It is clear that the rhetorical boundaries of science were constructed in such a way that transgressions were portrayed as nonscientific. University of Wisconsin nuclear engineer Tom Instrator argued that "that's not the way you do science … If they want [other scientists] to not give them a hard time, they should give us enough to copy it. Or maybe they have given us enough, and it doesn't work" (as cited in Allegretti, 1989).

Harold Furth of the influential Princeton Plasma Physics Laboratory testified that he was suspicious of the motives behind this secrecy, saying that "there is something here that he's [Fleischmann] not talking about, and I … look forward to some time when his patent attorneys or whoever will let him talk about it" (U. S. House of Representatives, 1989, p. 189).

Perhaps the most pointed deployment of this normative criterion appeared in a 1989 *New York Times* editorial by Lawrence Livermore National Laboratory physicist George Chapline. He insisted that "their behavior is deplorable. At any time in the past month, they could have submitted the electrodes in their apparatus for an independent analysis of their helium content. … Their failure to submit the electrodes for analysis leaves one wondering what is happening at the University of Utah" (p. 15). At the annual meeting of the American Physical Society in Baltimore, Caltech's Nathan Lewis expressed frustration over the inaccessibility of experimental details, bemoaning the fact that "Pons would never answer any of our questions" (Browne, 1989, p. 13).

Ironically, the constitutive authority attributed to intellectual communism was underscored by Pons and Fleischmann's acceptance of that norm as an appropriate standard of evaluation. (Of course, they rejected claims that their own behavior contravened the standard.) Responding to Congressional inquiries, Fleischmann defended their actions, insisting that "it is common practice to release a preliminary publication before you write a full paper. … We thought that we had given sufficient data in that preliminary publication that a cool and collected look at the paper would enable other people to replicate the experiment" (U. S. House of Representatives, 1989, p. 20). In response to further questioning regarding the withdrawal of their initial submission to *Nature,* he insisted that it "would not be the appropriate place to submit this work in the form of a full paper; they don't publish full papers" (U. S. House of Representatives, 1989, p. 21).

What is interesting here is not that Fleischmann differed with his critics on the proper interpretation of the Utah researchers' actions. That is to be expected. What is important is that widely disparate sets of practices could simultaneously be evaluated as in accordance with (or violation of) this cultural resource. This seems compelling evidence of the elasticity and open-endedness of at least this component of the rhetorical culture of science.

The significance of these scientific rhetorics does not stop at this broad communal level, however. This alleged violation of the normative order was directly linked to the epistemic status of experimental claims because the details were considered necessary for others to attempt possible replication. Full disclosure of details is thought to offer a hedge against unanticipated or unrecognized errors in experimental procedures. Finally, scientists linked this secrecy to the inability of most laboratories to confirm the Utah findings. An advisory board's final report concluded that the "investigators simply do not say in public reports exactly how the measurements are made; only by visiting the laboratories does it become evident that tabulations of quantities with so many [significant] figures is wholly unjustified. This practice is insidious … and it could explain why some groups see positive heat effects only erratically" (Energy Research Advisory Board, 1989, p. 48).

This rhetorical commitment was construed as sufficiently central to scientific practice that criticism of its violations even appeared in the technical literature. It is relatively uncommon for such statements to appear in technical reports because the structural constraints of journal publication actively eliminate personal references in favor of the appearance of an objective reality (Bazerman, 1987; Gilbert & Mulkay, 1982). For example, Kreysa, Marx, and Plieth (1989), detailing their efforts to replicate certain critical aspects of the Utah studies, indicated that "we recalculated the date of table 2 in ref. 1 [of the initial publication] which is not really easy since not all necessary data are given explicitly" (p. 443). Ziegler et al. (1989), reporting on their attempts to detect charged particles in a teflon electroanalytical cell, noted that "the above experiments are difficult to reproduce in detail since the authors discussed few details of the experimental technique" (p. 2929).

This general indictment of the secretive manner in which Pons and Fleischmann reported the background of their controversial experiments is given clearest expression in their avoidance of traditional peer review procedures. As I have noted, Pons and Fleischmann had announced their preliminary findings at the Utah press conference before their publication in (though not submission to) the *Journal of Electroanalytical Chemistry.* This was construed as a blatant violation of peer review. Ronald Ballinger of MIT testified that "for results such as those ... whose potential impact on the scientific community and the world are so great, this review is absolutely essential. Unfortunately ... this has not happened in this case ... And so we in the scientific community are left to attempt to reproduce or verify a potentially major scientific breakthrough while getting the experimental details from the Wall Street Journal" (U. S. House of Representatives, 1989, pp. 178–179).

The rhetorical authority, which is granted peer review, functioned as a regulatory discourse that sustained the community's position of epistemic privilege. As Frank Press, president of the National Academy of Sciences, saw it, peer review practices were central to the cultural authority of science. He suggested that "if the manner in which [Fleischmann and Pons] announced their findings is taken as a role model for others, the overall public view of science will be damaged ... The public will be disillusioned. The support for science will be eroded" (as cited in Detjen, 1989, p. D10). Arnold Rellman, then editor of *The New England Journal of Medicine* concluded that "the cold fusion story is a good case in point, even if it turns out to be true. The public doesn't know what to think. There's a lot of cynicism about science, a lot of time being wasted" (as cited in Strauss, 1989, p. 67).

BEYOND DISCIPLINARY RHETORICS

At this point, it should be apparent that disciplinary rhetoric was both constrained by and exerted transformative pressure on communal practices. However we ultimately evaluate the cogency of the Utah claims (and there seems to be considerable reason to doubt that cogency), and what is still unclear is how, precisely, disciplinary rhetoric was connected to the broader political practices to which I alluded earlier.

The connection, I think, is born of the capital-intensive nature of contemporary techno-science. Although the traditional image of the scientist puttering away in her or his laboratory, untouched by the vagaries of budgets, etc., provides grist for Hollywood and science textbooks, it bears little connection to contemporary techno-science. Bruno Latour (1987) aptly described the material ground of scientific practice, saying "When scientists appear to be fully independent, surrounded only by colleagues ... it means they are fully dependent. ... When they are really independent, they do not get the resources with which to equip a laboratory [or] to earn a living" (p. 158). The apparatuses that form the basis for so much of our advanced scientific research are made possible only through a relatively stable supply of research funds, whether from corporate interests or from the federal government via the DOE, NSF, and Defense Department, to name but a few. Put simply, science and scientists cannot "just do it." To an increasing extent, they resemble Tennessee

Williams's faded grande dame, Blanche Dubois, dependent "on the kindness of strangers."

Although kindness might have been less a consideration than technical viability, the requisite funds for cold fusion research were not forthcoming. University of Illinois materials scientist George Miley, filling the role of public advocate for the traditional fusion community, testified "I do not believe that cold fusion should force any near term changes in funding for the older established fusion research programs. We simply do not know enough about cold fusion yet. ... The basic physics [must] be firmly demonstrated" (U. S. House of Representatives, 1989, p. 146). The Energy Research Advisory Board rejected unequivocally any meaningful alteration of current funding practices. Their final report urged against "the establishment of special programs or research centers to develop cold fusion" (Energy Research Advisory Board, 1989, p. 1). It is significant to note that this recommendation assumed the primacy of the traditional commitments of the fusion physics community. The final report also indicated that "nuclear fusion at room temperature ... would require the invention of an entirely new nuclear process" (Energy Research Advisory Board, 1989, p. 3).

In this case, then, the decisions to deny funding for concerted cold fusion research programs effectively doomed their future development, whatever the ontological status of cold fusion. Without the requisite materials with which to expand the experimental base and, in the (however unlikely) event of positive results, with which to produce energy efficiency, the cold fusion program could not possibly succeed. As Peat (1989) put it, "The bottom line on fusion power is a matter of economics. Fusion laboratories—whether they be tokamaks, magnetic mirrors, or lasers firing at frozen pellets—are megaprojects and require megadollars" (p. 36).

Although the Utah team's request for approximately $50 million pales by comparison to budgets for projects like the SSC, those so-called mega-dollars are at the heart of laboratory practices. Brigham Young University physicist Daniel Decker, calling for maintenance of the current funding structure, offered a direct illustration of the interaction of the rhetorical and material dimensions of the controversy. He maintained that "in order to see the fusion, very sensitive neutron detectors are required which are not available in most laboratories. These are not expensive, but they are rare. ... I would think 10 to 20 million dollars ... would be adequate" (U. S. House of Representatives, 1989, p. 126). That the detectors are not expensive (by the standards of big science) is of little consequence. Whatever the cost, the key issues are that they are prerequisites to knowledge production and that their acquisition is dependent largely on rhetorical transactions occurring outside the laboratory.

Accordingly, the ostensibly political deliberations that yielded the judgment to deny funding should not be written off as nonscientific, or even as only marginally scientific. It seems to me, rather, that the conduct of laboratory practice was directly and inherently tied to their outcomes. This is not to say that Robert Roe, chair of the Committee on Science, Space, and Technology, donned a lab coat and measured neutron flux. It is to say that laboratory practice is but one constituent of the cultural (and rhetorical) practices of science in the modern age. As the *Congressional Quarterly*

Weekly Report concluded, "Cold fusion ... is a ... political phenomenon ... subject to the stratagems, tradeoffs and turf wars that accompany politics—and Congress is right in the thick of it" (Cloud, 1989, p. 960).

CONCLUSIONS

The orientation adopted in this chapter suggests that a further reconceptualization of the basic analytic categories of social (public) and technical might be in order. Traditional conceptions of these "spheres of discourse" (Goodnight, 1982) assume (with good reason) that technical rhetorics have progressively eroded the vitality of debate in the public sphere. This study suggests that, at least in principle, we need not despair in the face of technical rhetorics. If my analysis is sound, it follows that practices typically thought to fall in the public sphere can exert directional influence on the corrosive technocracy that many rhetorical critics have decried (e.g., Farrell & Goodnight, 1981). I do not suggest that, somehow, all rhetorics are really all the same, that differences in content and context are chimerical. My point here is only that what we have traditionally thought of as nonscientific rhetorics can, in many contexts, be read (and criticized) as among the important cultural practices of science.

The notion of a scientific ecosystem might be worth considering. The constituents of our natural ecosystem exist in a fluctuating equilibrium. Tipping the balance in one direction or another will send reverberations throughout the system. Systemic imbalances in favor of Pacific Northwest timber interests or of Earth First! would irrevocably alter the fragile ecosystem to the profound detriment of the larger whole. The constituents of the scientific ecosystem (understood as the complex network of cultural practices) would function similarly. Tipping the balance solely in favor of "science in its own terms" would point toward a corrosive technocracy. On the other hand, opting for government intervention in technical decisions raises the spectre of Lysenkoism and a corresponding degradation of intellectual rigor, or worse. To the extent that we understand science as having nontrivial interfaces with political practices, we are presented the opportunity to craft meaningful interventions into traditionally technical matters from the vantage points of rhetorical critics and public advocates.

Of course, analysis of such discourses as those issuing from congressional hearings, funding deliberations, and even from journalists as "scientific rhetorics" will no doubt make some uncomfortable. It could be said that not all such discourses are really scientific, that perhaps I have defined the relevant cultural practices so broadly that science (as any sort of practice) has lost any discrete identity. This is a serious challenge and one that warrants response.

In the first instance, I would note that I do not ordain in advance which rhetorics are scientific, nor do I say that all practices are scientific in nature. Unless and until a preponderance of scholarship lead us to legitimize only a limited set of social practices as uniquely definitive of science, I think the heuristic value of casting our theoretical net widely is preferable to insisting on a particular route before we have decided where we are going.

A second reason to reject (or at least defer) the assumption that science is fundamentally a homogeneous phenomenon is that it carries the related assumption that science, whatever it is, comes complete with its own interpretive repertoire. Although practicing scientists certainly have their own routines and methods of accounting for them, that need not entail that all interpretations of said practices need not be isomorphic with them. Indeed, we now recognize that most intentional scientific rhetorics attempt to mask their own rhetoricity (Gilbert & Mulkay, 1982, pp. 39–111). The facts of science cannot speak for themselves; they must be spoken for.

If we want to sustain both a rigorous approach to the analysis of scientific rhetorics and to our traditional concern with promoting a more humane division of intellectual labor, we must avoid being trapped within precisely the disciplinary rhetorics we take as our objects of study (Leff, 1987). Taking rhetorics of science solely on their own terms (of self-denial) can do little except to reinvigorate the reigning power/knowledge hierarchies which so rightly trouble us. Perhaps Herbert Marcuse (1960) had it right when he praised the "power of negative thinking," saying it could "undermine the sinister confidence in the power and language of facts" because "to express and define that which is on its own terms is to distort and falsify reality" (p. x).

By investigating the myriad points at which these rhetorics converge to produce science in contemporary culture, critics can illuminate the scientificity of modern life and the rhetoricity of modern science. By so doing, we can begin to craft more inclusive and humane visions of science and culture, while rejecting the "exoticism of the other" (Woolgar, 1988, p. 108). Who or what is constituted as "other" in the discourses of science is a contextual production, and critical vigilance of those acts of production is a fundamental characteristic of this theoretical perspective.

Although this view provides one theoretical framework, it certainly does not dictate the contours of research conducted within that framework. Evaluations of the case study offered here and any subsequent work remain the product of the critical conversation. Our conversants do, in fact, make evaluations of the quality, legitimacy, accuracy, etc., of the various contributions to the conversation. The currency of the conversation, thus, must be argumentative (Brockriede, 1974). It is this insistence on open-ended argumentation that Michael Billig (1989) sees as the key to avoiding rhetorical conservatism and enabling meaningful critique. The implications of such a position are aptly captured in his retelling of the Talmudic parable which concludes "The law is not in the heavens. The law, indeed all laws, lay in earthly discussions. If heavenly voices wished to join in the debates, they must do better than shake walls or send carobs flying: they must present arguments" (Billig, 1987, p. 29).

The critical jury is still out on the theoretical orientation for the critical analysis of scientific rhetorics developed here. Construing science as a series of heterogeneous cultural practices allows rhetorical critics to make significant theoretical contributions to the interdisciplinary conversation of science studies. It is the process of theoretical demarcation I have attempted here that will, I hope, allow us to avoid the charges of theoretical disarray or even sterility others have lodged against rhetorical studies of science (Gross, 1992; Myers, 1989; Myers, 1990).

REFERENCES

Allegretti, D. (1989, April 20). Fusion claim puzzles UW profs. *Capital Times [Madison WI]*, p. F9.

Bazerman, C. (1987). Codifying the social scientific style: The APA publication manual as a behaviorist rhetoric. In J. Nelson, A. Megill, & D. McCloskey (Eds.), *The rhetoric of the human sciences: Language and argument in scholarship and public affairs* (pp. 125–144). Madison, WI: University of Wisconsin Press.

Billig, M. (1987). *Arguing and thinking: A rhetorical approach to social psychology*. Cambridge: Cambridge University Press.

Billig, M. (1989). Conservatism and the rhetoric of rhetoric. *Economy and Society, 18,* 132–148.

Bishop, J., & Wells, K. (1989, March 24) Taming H-bombs? Two scientists claim breakthrough in quest for fusion energy. *Wall Street Journal,* p. 1.

Brockriede, W. (1974). Rhetorical criticism as argument. *Quarterly Journal of Speech 60,* 165–174.

Browne, M. (1989, May 1). Fusion claim is greeted with scorn by physicists. *New York Times,* p. 13.

Campbell, J. (1986). Scientific revolution and the grammar of culture. *Quarterly Journal of Speech, 72,* 351–376.

Chapline, G. (1989, July 25). Cold water on cold fusion. *Chicago Tribune,* p. 15.

Close, F. (1991). *Too hot to handle: The race for cold fusion*. Princeton, NJ: Princeton University Press.

Cloud, D. (1989). Utah's fusion scheme may create fission on hill. *Congressional Quarterly Weekly Report, 47,* 960.

Collins, R., & Restivo, S. (1983). Development, diversity, and conflict in the sociology of science. *The Sociological Quarterly, 24,* 185–200.

Detjen, J. (1989, April 30). Scientists fearing fallout from cold fusion controversy. *Philadelphia Inquirer,* p. D10.

Energy Research Advisory Board. U. S. Department of Energy. (1989). *Final report: Cold fusion research* (DOE/s-0073). Washington, DC: Author.

Farrell, T., & Goodnight, G. T. (1981). Accidental rhetoric: The root metaphors of Three Mile Island. *Communication Monographs, 48,* 271–300.

Fleischmann, M., & Pons, B. S. (1989). Electroanalytically induced nuclear fusion of deuterium. *Journal of Electroanalytical Chemistry and Interfacial Electrochemistry, 261,* 301–308.

Fleischmann, M., Pons, B. S, & Hawkins, M. (1989). Erratum. *Journal of Electroanalytical Chemistry and Interfacial Electrochemistry, 263,* 187–188.

Fleischmann, M., Pons, B. S., & Hoffman, R. (1989). Measurement of y-rays from cold fusion. *Nature, 339,* 667.

Gieryn, T. (1983). Boundary work and the demarcation of science from non-science: Strains and interests in professional ideologies of scientists. *American Sociological Review, 48,* 781–795.

Gilbert, G. N., & Mulkay, M. (1982). Accounting for error: How scientists construct their social world when they account for correct and incorrect belief. *Sociology, 19,* 165–185.

Gilbert, G. N., & Mulkay, M. (1984). *Opening Pandora's box: A sociological analysis of scientists' discourse.* Cambridge: Cambridge University Press.

Goodnight, G. T. (1982). The personal, technical, and public spheres of argument: A speculative inquiry into the art of public deliberation. *Journal of the American Forensic Association, 18,* 129–144.

Greenberg, D. (1989, May 9). Turf and money—The buttons cold fusion pushed. *Chicago Tribune,* p. 17.

Gross, A. (1991). Rhetoric of science without constraints. *Rhetorica, 9,* 283–299.

Gross, A. (1992). Call for Papers. *Technical Communication Quarterly, 1*(1), 71.

Holmquest, A. (1990). The rhetorical strategy of boundary work. *Argumentation, 4,* 235-258.

Hynes, T. (1989). Can you buy cold fusion by the six pack? Or, Bubba and Billy Bob discover Pons and Fleischmann. In B. Gronbeck (Ed.), *Spheres of argument: Proceedings of the Sixth SCA/AFA Conference on Argumentation* (pp. 42–46). Annandale: Speech Communication Association.

Kreysa, G., Marx, G., & Plieth, W. (1989). A critical analysis of electrochemical nuclear fusion experiments. *Journal of Electroanalytical Chemistry and Interfacial Electrochemistry, 266,* 437–450.

Latour, B. (1987). *Science in action: How to follow scientists and engineers through society.* Cambridge, MA: Harvard University Press.

Leff, M. (1987). Modern sophistic and the unity of rhetoric. In J. Nelson, A. Megill, & D. McCloskey (Eds.), *The rhetoric of the human sciences: Language and argument in scholarship and public affairs* (pp. 19–37). Madison, WI: University of Wisconsin Press.

Lyne, J. (1990). Bio-rhetorics: Moralizing the life sciences. In H. Simons (Ed.), *The rhetorical turn: Invention and persuasion in the conduct of inquiry* (pp. 35-57). Chicago: University of Chicago Press.

Lyne, J., & Howe, H. (1990). The rhetoric of expertise: E. O. Wilson and sociobiology. *Quarterly Journal of Speech, 76,* 134–151.

Mallove, E. (1991). *Fire from ice: Searching for truth behind the cold fusion furor.* New York: Wiley.

Marcuse, H. (1960). *Reason and revolution.* Boston: Beacon.

McAllister, J. (1992). Competition among scientific disciplines in cold nuclear fusion research. *Science in Context, 5,* 17–50.

McCann, H. (1989, May 12). Did the fusion pair hide essential detail to protect patent claim? *Detroit News,* p. F11.

McGuire, J. E., & Melia, T. (1991). The rhetoric of the radical rhetoric of science. *Rhetorica, 9,* 300–317.

Merton, R. (1973). *The sociology of science: Theoretical and empirical investigations.* Chicago: University of Chicago Press.

Miller, C. (1989). Some perspectives on rhetoric, science, and history. *Rhetorica, 7,* 101–114.

Myers, G. (1989). Persuasion, power, and the conversational model. *Economy and Society, 18,* 220–244.

Myers, G. (1990). Sociology of science without the sociology. [Review of the book *A rhetoric of science*]. *Social Studies of Science, 20,* 559–563.

Overington, M. (1977). The scientific community as audience: Toward a rhetorical analysis of science. *Philosophy and Rhetoric, 10,* 143–164.

Peat, F. (1989). *Cold fusion: The making of a scientific controversy.* Chicago: Contemporary.

Pickering, A. (1992). From science as knowledge to science as practice. In A. Pickering (Ed.), *Science as practice and culture* (pp. 1–26). Chicago: University of Chicago Press.

Prelli, L. (1989). *A rhetoric of science: Inventing scientific discourse.* Columbia, SC: University of South Carolina Press.

Rothman, M. (1990). Cold fusion: A case history in "wishful science"? *Skeptical Inquirer, 14,* 161–170.

Schmidt, W. (1989, April 18). Other laboratories back findings, chemist says. *New York Times,* p. 28.

Speech Communication Association. (1992). Special *Technical Communication Quarterly* Issue on "Rhetoric of science and technology." *Spectra, 28.*

Strauss, H. (1989, April 16). Ga. Tech learns risks of going public too soon. *Atlanta Journal,* p. G7.

Taylor, C. (1991a). Defining the scientific community: A rhetorical perspective on demarcation. *Communication Monographs, 58,* 402–420.

Taylor, C. (1991b). The public closure of technical controversy. In D. Parson (Ed.), *Argument and critical practices: Proceedings of the Seventh SCA/AFA Conference on Argumentation* (pp. 263-269). Annandale: Speech Communication Association.

U. S. House of Representatives. (1989, April 26). *Recent developments in fusion energy research.* (Committee on Science, Space, and Technology Publication No. 26). Washington, DC: Author.

Woolgar, S. (1988). *Science: The very idea.* Chichester: Ellis Horwood.

Young, L. (1989, May 2). Academia's heavies weigh in, pour cold water on Utah fusion theory. *Baltimore Sun,* p. E10.

Ziegler, J. F., Zabel, T. H., Cuomo, J. J., Brusic, V. A., Cargill III, G. S., O'Sullivan, E. J., & Marwick A. D. (1989). Electrochemical experiments in cold fusion. *Physical Review Letters, 62,* 2929–2932.

part III

The History of Scientific Discourse

7

"Practicing Our Preaching": Gifford Pinchot, Conservation, and the Scientific Rhetoric of "Demonstration Forestry"

J. Harrison Carpenter
Michigan Technological University

The historian Samuel Hays's landmark work *Conservation and the Gospel of Efficiency* (1959) challenges the widely held idea that the American conservation movement of the late 19th century was a struggle between greedy, "evil 'interests'" and their prey, the "virtuous 'people.'" Such a simple dichotomy, he claims, neglects the often contradictory positions of both these groups on conservation issues and creates an overly reductive picture of the complex social situation that germinated the movement toward conservation policy. Hays further explains that the early conservation movement "neither arose from a broad popular outcry, nor centered its fire primarily upon the private corporation." He thus calls for a rethinking of conservation history, after which "an entirely new frame of reference" that fully considers the social text will "arise from the evidence itself" (pp. 1–2).

For Hays (1959), the central factor in developing the conservation impulse and mindset as a social reality was the development and proliferation of American applied science: "Conservation, above all, was a scientific movement, and its role in history arises from the implications of science and technology in modern society" (p. 2). Since it was led by scientists, conservation necessarily reflected their disciplinary and ideological biases and preferences. Fervently believing that natural processes and relationships were essentially technical matters, the conservation leaders shared the position that resource policy decisions were rightly made by technicians, more so than by legislators, and least of all by the emotional and unknowledgeable public. They envisioned a political system built along scientific lines, where these technical decisions were made with the "cold rationality" of science and not by the heated exchange of political rhetoric.

Yet simply to declare conservation a scientific endeavor is an incomplete statement. Although many conservation leaders were trained in the sciences (some, like Theodore Roosevelt, were not), Hays (1959) admits the obvious difficulty of equating the views of "a few" scientists with a broad-based public issue like conservation (p. 3). Beyond this concern, calling conservation an exclusively scientific movement is dangerously close to the kind of reductionism Hays specifically challenges his reader to avoid; it keeps the binary distinction between rhetorical agents intact, but it substitutes scientists for the "people" and the forces of irrationality for the "interests." Further still, although Hays strongly emphasizes that we must view conservation "from the vantage point of applied science, rather than of democratic protest" (p. 2), he offers his readers little opportunity to understand the movement's leaders *as scientists.* If we are truly to see conservation as a scientific movement, we must consider it as constituted in other scientific practices. To do so requires a deeper discussion of scientific practice than can reasonably be offered in a historiography. This chapter propounds to extend Hays' formulation, by way of examining the typical scientific practice of report writing. I conceive of this practice as a rhetorical enterprise through which scientists craft "practical knowledge, a vehicle for practical reasoning" (Gross, 1990, p. 5).

PINCHOT AND THE INFLUENCE OF BERNHARD FERNOW

I have chosen as representative a pair of reports authored by one of conservation's leading proponents, Gifford Pinchot, under whose supervision the U.S. Forest Service and National Forest system took on its current form. Pinchot, the great "popularizer" of the principles of Progressive conservation, was not the first leader of American governmental forest administration. Agents of the government had inquired into the plight of American forests since the late 1860s, but professional forestry was not to have an official presence until 1876, when the position of Confidential Forest Agent was added to the U.S. Department of Agriculture. By 1881, the Agent was overseeing a small staff of assistants in what was renamed the Division of Forestry (Steen, 1976).

In 1886, Bernhard Fernow—a botanist, naturalist, and forester licensed in his native Germany—took the head of the Division. It was Fernow (under whom Pinchot began his professional career in 1892) who acquainted Pinchot with the views of naturalist George Perkins Marsh. Marsh argued that past civilizations had fallen due to the wasteful use of natural resources (1965, pp. 9–13). Fernow (1886) advocated that foresters take a proactive direction by providing the public with examples of "systematic and successful management" of select forests, a program he called "demonstration forestry" (p. 166). Despite his conviction, Fernow was unable to muster wide support for his ideas. Like other European forest advocates, his arguments were based primarily on the idea that a "timber famine" (a time when marketable logs were in drastically short supply) was imminent. This was a contention very much open to attack from timber interests and others who doubted that anyone could accurately estimate American timber volume and depletion rates (Clary, 1986, pp. 18–19). One lumber agent, calling scientific foresters "unhappy dreamers," asserted: "Some of the purely theoretical [forestry] may be of use, but the average mill owner has had considerable experience with the holder of theoretical ideas ... and looks [upon him] with suspicion and disfavor" (as cited in Clary, p. 18).

By the time Pinchot succeeded Fernow in 1898, Pinchot's belief in the viability of demonstration forestry was solidified; however, Pinchot did not share Fernow's tendency toward gradualism. Feeling that "the thorough and intensive methods used in the old world forests had no relevance in the American environment," Pinchot was determined to "take management directly into the woods and to learn his methods empirically as he went" (Dupree, 1985, pp. 240–241), an approach that he called "practical forestry" (Pinchot, 1891). Pinchot's unqualified success at converting the discipline to his vision of practical forestry is indicated by the strong *esprit de corps* that infuses the modern U.S. Forest Service. A further indication is the strong line of separation that is present between applied "forest management" and experimental "wood science." In inculcating forestry as an applied science, Pinchot made it clear that the discipline was primarily a technical field, operating under a purposive rationality that "considers ends and means, and the correct calculation of methods to achieve results" (Myerson, 1994, p. 21).

As the avowed leader of the conservation and forestry causes, Pinchot has been recognized by numerous historians (for example, McGeary, 1960; Miller, 1994; Pinkett, 1970) as occupying a pivotal role in shaping American forest policy. He has also been noted by a number of rhetoricians (Oravec, 1984; Killingsworth & Palmer, 1992; Meisner, 1995) as one of the preeminent figures of environmental rhetoric. However, few rhetorical studies have attended to Pinchot's primary works in systematic analytical depth. Yet these works do hold much promise for those who study the history of environmental rhetoric and the discourse of applied science. Although small points in Pinchot's long and noteworthy career, his reports on demonstration forests occupy a unique position in the history of environmental rhetoric. Released to the public, the documents of demonstration forestry hallmark the first attempts to win support for the conservation cause by appealing to nonaesthetic values. As such, Pinchot's reports are not merely scientific documents, but the first instances of a genre that today is the privileged paradigm in government environmental discourse. Furthermore, these reports represent this early leader's successful attempts to define forestry practices at a time when that discipline was still in its formative stages. More generally, a rhetorical analysis of these reports—undertaken with an awareness of their status as artifacts in the development of forest science and in environmental history—suggests that scientific discourse can serve both persuasive (deliberative) and affirmative (epideictic) rhetorical goals given certain situational factors (exigences). Examining these rhetorical concerns in Pinchot's demonstration forestry campaigns can very well deepen our awareness of conservation history as it has developed with (and through) science.

RHETORIC AND GIFFORD PINCHOT'S FORESTRY

Scholars of rhetoric have in the past focused on a narrow range of rhetorical issues in conservation history. Killingsworth and Palmer (1992) offer one such comment. Building on Hays's analysis, they claim the decisions made by Pinchot and other leading conservationists began a long string of events whose consequences are ultimately antidemocratic. Arguing that "the advent of scientific management in American government was accompanied by a devaluation of the potential contribution of local governments and the general public to national policy," Killingsworth and Palmer assert that for conservation's "scientifically trained experts, for whom Pinchot himself provided the model," public opinion was "to be neutralized, not incorporated into the decision-making process" (pp. 164–165).

To understand conservation as a scientific movement behooves us to move away from the rhetoric of protest into the rhetoric of science; that is, the process by which scientists make "warranted assertions about contingent claims that are based on 'good reasons'" (Weimer as cited in Prelli, 1989, p. 7) so as to make legitimate claims to knowledge and community membership. As a rhetorical product, "scientific discourse is accepted or rejected on grounds of its *reasonableness*—given the issue at stake, the knowledge conditions of the scientific community, and the perceived expertise of the makers of the claims" (Prelli, p. 7). In the case of Pinchot's demonstrative reports, however, there are additional concerns that warrant a slightly different brand of consideration. Foregrounded among these issues are the documents' public nature. Traditionally, rhetoricians of science have examined a discourse to throw light upon: "how it is judged as *science*" (Prelli, p. 8) and "the efforts made by scientists to persuade one another" (Wander, 1976, p. 227). Any systematic study of Pinchot's rhetoric must also take into account the more complex situation that results when scientific documents are taken directly into the public arena.

Furthermore, when assessing the rhetoric of demonstration forestry, one must also take into account the scientific milieu in which Pinchot was immersed. By the mid-1890s, forestry in the U.S. was an immature but developing field whose primary supporters were the botanists, biologists, and geologists who had reason enough to study silviculture (Clary, 1986, p. 6). Many of the leading voices in forest conservation—such as John Muir, the celebrated nature writer—argued for the transcendental or aesthetic value of wilderness and not for the forest's potential as an object of rational management or scientific study (Oelschlaeger, 1991, pp. 173–175). What literature there was on forestry was European; the works of German *forstmeister* Sir William

Schlich were canonical and influenced American foresters toward conservative harvesting. Clary (1986) explains:

> European forestry, developed in countries that had long been occupied and in which populations were growing but native resource bases were not, husbanded timber resources for future use. It therefore averted the threat of a "timber famine" by applying strict technical and political purposes. (p. 6)

American foresters soon surmised, however, that the practices of Europe were not directly translatable to the U.S. A considerable amount of effort would have to be made to define and perfect the discipline's domain on the American continent, just as it had to be in Europe beforehand. In the conversations about practice, Pinchot's rhetoric was to weigh heavily, as he sought to codify and delimit the young profession's bailiwick.

Given this situation, one is called upon to consider Pinchot's scientific rhetoric as simultaneously forging the standards of his nascent discipline, as well as securing government and private support for forest management and conservation more generally. For Pinchot, then, the wisest rhetorical path was at once formatory and reconstitutive. In shaping his demonstrative reports, Pinchot not only had to adhere to the usual standards of scientific confirmation (Gross, 1990): Owing to the complexities of the situation, Pinchot also had to gain the assent of other audiences (notably landowners and legislators, upon whose support the Bureau of Forestry depended) by advancing a reasonable view of the forest not based on aesthetic or mystical grounds. Additionally, Pinchot was called upon to establish practical forestry as a paradigm and demonstrate the validity of forest management practices in the face of prevailing skepticism. Just how Pinchot was able to accomplish these goals is the topic to which I now turn.

BILTMORE FOREST AND THE RATIONAL RECONCEPTION OF FORESTS

Pinchot actually began his demonstration forestry activity before heading the Division of Forestry; in 1892, fresh from his studies abroad, Pinchot was appointed to manage the woodlands of George W. Vanderbilt's sprawling Biltmore Estate. After a year of work, and in cooperation with the Division, Pinchot published *Biltmore Forest*, a brief booklet that was distributed at the Forests exhibit hall during the 1892–1893 Columbian Exhibition in Chicago. *Biltmore Forest* detailed Pinchot's attempts to "treat Biltmore Forest systematically," which he confidently labeled "the first practical application of forest management in the United States" (1893/1970a, p. 5). Although meant primarily to spread the gospel of forest management, *Biltmore Forest* made way for a new understanding of forests, based not on aesthetics or mysticism, but on a plainly empirical accord. For instance, the document contains large amounts of numerical data taken from firsthand observation. Measurements are given to the nearest whole number for area (p. 8); meteorological data is displayed to one or in some cases two decimals (p. 9). An impressive list of the forests' species of trees—more than 60—takes up almost three pages of the text (pp. 11–13).

Such devices connote the detail and care with which practical forestry is carried out. These figures' effect is bolstered by a direct and objective writing style that often states content in a way that appears inarguable:

> The summits of the hills, which reach an average altitude of about 2,300 feet, mark the approximate height of the Asheville base-level, from which, through erosion, the present outline of the surface was derived. (Pinchot, 1893/1970a, p. 8)

> [The Forest] is composed in greater part of Oaks and other deciduous trees, chiefly in the younger stages of their growth, with Pines scattered among the broad-leafed trees, and here and there patches of the old fields. (p. 8)

> During the winter the relative humidity is occasionally high, but this is generally not the case at other seasons. … The relative humidity, taken by monthly averages, was least in May (54.91%), and greatest in January (76.01%). (p. 9)

Notice that only in this last sample do we find a hint of human agency ("taken by monthly averages"). In *Biltmore Forest*, Pinchot the observer is invisible; the facts of forestry come to us *qua* facts. Pinchot's style reflects the effusive discourse of his time, but highlighted especially are devices that demonstrate the systematic and rigorous nature of the study. By association largely, Pinchot the researcher appears to be both professional and efficient, using his evident knowledge to draw decisive conclusions and to act confidently. On a more theoretical level, such a style carries with it several assumptions about the separation of facts and values that cannot go unnoticed.

First, Pinchot argues implicitly for his inclusion in a community separate from his audience, and the knowledge of that community is internal. Though Pinchot suppresses in his discussion any idea that human perception and choice shapes the study of the forest, even a cursory examination of these and other passages reveals numerous decisions that were made, decisions that both reflect and affect Pinchot's conception of forestry and forests in general. For instance, Pinchot mentions temperature, humidity, and rainfall as factors of climate; why were these measurements chosen to describe the forest habitat? How has he determined that the forest "lies near the centre of distribution of the forest flora of Eastern North America" (Pinchot, 1893/1970a, p. 11)? How has it been shown that erosion affected the "Asheville base-level"? By denying his readers answers to these and other questions, Pinchot makes judgments about what is and is not appropriate information for his readers, assuring them a subordinate place. They must go to the foresters for more information.

Second, Pinchot separates the forest ecosystem into components: Climate is separate from soil, which is separate from the trees, which are distinct individuals unto themselves. These elements bear no necessary relation to each other, but come together to make a woodland. Animals are conspicuously absent, and the only plant species mentioned are woody; this partially represents Pinchot's own preoccupations and limitations. As a forester, he was obliged to think of trees first; yet the atomistic quality of the description cannot be denied. Pinchot (1893/1970a) lists the species present (pp. 11–12) but does not approach their relation to each other, nor does he offer any coherent explanation of the forest's development and propagation. For example, he refers to Biltmore as being "composed for the most part of deciduous trees" (p. 10); note that this characterization does not make apparent the forest *as a whole.* Rather, the forest is equal to its components: numbers of individual "deciduous trees." The ease with which Pinchot separates individuals from their supportive networks must have reassured his readers that selective cutting would not disturb the integrity of the forest.

Finally, removing the human aspects from his descriptions makes it easier for Pinchot to argue the value of forestry practices on rational grounds. Pinchot (1893/1970a) presents his plans and describes his actions not as conscious attempts at management and control, but as "intelligent supervision" that allows "improving the forest" to stand paramount (p. 31). A primary imperative of this practice was the saving of young trees from premature death: "The old spreading trees were seriously injuring the young growth below them. … It became necessary, therefore, to institute a series of improvement cuttings which should remove these older trees" (p. 18). The forester appears here not as a manipulator, but as an equalizer in Nature's plan; older trees are smothering their progeny, so to speak, and are sapping the forest of its health and vigor. Rather than wait for the older trees to die (many "were already perishing"), the forester steps in to assure the next generation's prosperity.

This not only reaffirms the rightness of forestry's practices, but also presents the forest as a hostile place even for its own denizens. Trees are in ruthless competition with each other for scarce resources, and only the strong will survive. To assure the survival of the forest, the forester intervenes, redistributing the forest's resources by removing the old growth that prevents young trees from getting their share. This view of competition within the forest and the equitable distribution of scarce resources is more directly evident in Pinchot's later writings on forests and forest health (Pinchot, 1899; Pinchot, 1905).

Biltmore Forest makes plain that "practical" forestry is a *technical* activity, whose practices are constituted in precise measurements and calculations as well as painstaking field study. This was in contrast to the aesthetic study of forests and individual species that had dominated American "forestry" up to this time (Hays, 1959, p. 27). From *Biltmore Forest* on, Pinchot advanced strong appeals to the empirical study of forests. By association, he advanced appeals to the authority of professional foresters, under whose management such studies were most efficiently and intelligently carried out. Hence, in documenting his early demonstration projects, Pinchot described in vivid detail the actions of his crews, often using figures to further emphasize the exacting nature of the work. Pinchot (1893/1970a) first used this device in *Biltmore Forest*: "Each [species] description was written by means of abbreviations on a card of the size of the old postal-card (3 x 5 inches); and the whole series, something over 600 in number, was arranged in the form of a card catalogue" (p. 22). He was to continue to use it in later documents, signaling both the increasing confidence Pinchot had in his methods and the growing systematization of his practical forestry paradigm.

DEFINING A PROFESSIONAL PURVIEW: *THE ADIRONDACK SPRUCE* AND "INTELLIGENT SUPERVISION"

In an 1898 report, *The Adirondack Spruce*, Pinchot (1898/1970b) describes the general purview of his work in New York's Adirondack Park: "The field work done comprised the counting and measurement, of 1046 acres, of all Spruces down to diameter of two inches, and of all other trees ten inches or more in diameter" (p. 3). Such minute details abound in this text and offer a thorough (if not overly enumerated) picture of foresters' work. Pinchot was also to give the "reasons for forest management," such that his audience could come to an understanding of foresters' intent as well (pp. 102–105; see also 1893/1970a, pp. 43, 49).

Early on, Pinchot had sought to distance himself from the aesthetic preservationists by subtly changing the rhetoric of forest conservation. In the documents of demonstration forestry, this often took the form of ethical appeals to an unnamed forester, one that presumably stood for all foresters and certainly for Pinchot himself. In *The Adirondack Spruce*, Pinchot not only reports on his management activity, but also clarifies and delineates the place of forestry in conservation and in the lumber industry. The role of "the forester," Pinchot (1898/1970b) explained, was not to prevent landowners from realizing a profit, but rather assist them in achieving better returns on their investments (p. iii). Pinchot's rhetoric stakes out a new territory for professional forestry, one based not on opposition to lumbering, but in relation to it:

> Perhaps nothing has been done more to retard the progress of forestry in American than the disregard of its intimate and friendly relation to lumbering—a relation which was almost wholly overlooked for years. … In the eyes of many of its early friends the lumberman was a vandal whose inordinate greed called for constant denunciation, while to the lumberman the ideas of the forest reformer had no relation whatever to the affairs of practical life … lumberman and forester are as needful to each other as the ax and its helve. Without the ax the helve has little weight; without the helve the ax is lacking both in reach and in direction. (pp. iii–iv)

Here, we see not only Pinchot's attempts to discredit the preservationists directly, but also an attempt to bolster his own position as an "intimate friend" of lumbermen. Pinchot's metaphor of the ax and helve must have been quite easy for his audience to grasp. The characterization of each part is precise. If the ax is taken to be lumbering, and forestry the helve, it is suggestive of timber harvesting's shaky defensive posture against preservationist's aesthetic arguments; with the added "weight" of technical expertise behind lumbering, it can be defended as a science. What is more, forestry is dependent upon lumbermen for its "reach and direction," that is, for its implementation and operation.

It is also with *The Adirondack Spruce* that Pinchot was to clearly separate his practical forestry from Fernow's intellectual approach. Although sticking to the intensive and precise detail of earlier reports, *The Adirondack Spruce* contains only 30 pages of nomenclature and climatic observations, compared to 116 pages of volume tables, returns, and working plan outlines. A full 49 pages are devoted to discussion of the working plan alone; clearly, this signals an attempt to shift consideration away from observation toward planning and rational management. In fact, Pinchot (1898/1970b) makes this point early in the document:

> The purpose of the present study was to acquire such a knowledge of the laws which govern the growth of the Red Spruce that the main facts necessary for the conservative treatment of Spruce forests might be put within reach of the men most likely to undertake it. (p. 1)

In a move typical of scientific writing, Pinchot here outlines the scope of his study and in so doing, circumscribes his area of concern. The fact that he requires only "the main facts necessary" again reiterates the appeal of efficiency, and also suggests the applied nature of the practical forestry paradigm. If his readers accepted Pinchot's promise to put forestry "within the reach of the men most likely to undertake it," then the appeal of an applied approach is obvious over a theoretical one.

Pinchot's attempts at defining an applied forestry paradigm in *The Adirondack Spruce* are bolstered by brief, but telling, descriptions of professional forestry's purview and philosophy. When discussing "The Basis for Forest Management" (1898/1970b, p. 34), Pinchot begins first by establishing a link with his audience through demonstrating his support for their common interest—lumbering. His tone at this point is very deferential: "It would be unfair to condemn all lumbering because the present method is not wholly right" (p. 34). The precisely used phrase, "the present method is not *wholly* right," is at once affirmative of the status quo and subtly challenging; Pinchot reassures his audience that he does not desire to "condemn all lumbering," merely change those parts of the practice that are flawed.

He then continues, weaving the concerns of foresters and lumber interests:

> It is the purpose of practical forestry so to modify the present systems of cutting that the harm now done may be avoided and the removal of the old crop may be of permanent benefit to the forest which remains. ... Better methods of lumbering affect the forest chiefly through the kind and amount of the reproduction and growth which follow cutting, and through the conditions under which trees developed to marketable size. (Pinchot, 1898/1970b, p. 34)

Pinchot here skillfully weaves *pathos* with *logos* to justify his actions. The practical appeal of forest management—protecting the future "crop" through careful "removal," and the supervising of "reproduction and growth" of that future crop—is interwoven with an appeal to the more abstract value of forest health. The "permanent benefit" of practical forestry goes not to the landowner, but to the "forest which remains" after cutting; "the protection of young trees" and "the encouragement of new growth" (p. 37) are among the positive outcomes for the forest itself. Through thinning, Pinchot claims, the remaining trees will be better examples of their kind; again, forester and lumberman appear as the ones who will save the forest from the damaging forces within:

> When a piece of ground is full covered with forest trees ... a struggle follows in which some trees are suppressed and checked in their development ... when a crowded stand is thinned the trees which remain grow more rapidly than before ... The practice of thinning is based on this capacity for increased growth on the part of the trees which have been more or less vigorously set free. (pp. 37–38)

Again, in the passage above, Pinchot posits the forester as an equalizer, assuring that every individual of the species at hand is given the opportunity to prosper. The parallel with the political discourse of Theodore Roosevelt's Square Deal and the Darwinistic narratives of the progressive era is apparent.

While reiterating foresters' ability to manipulate forest ecology forthrightly, Pinchot is able to justify their actions by appealing to a value beyond question: the forest itself. Such a construction obscures the value judgments inherent within it, as Latour and Woolgar (1979) have concluded, and has as its effect "the assertion or implication that what is said and done is all that *could* be said and done within the research situation" (Prelli, 1989, p. 103; original emphasis). Pinchot (1898/1970b) then connects the concern for forest health to an egocentric appeal to the landowners' self-interest: "At the same time the question of revenue is kept prominently in view ... [better lumbering methods] affect the owner through the sustained and increased value of his property and its yield" (p. 34); "Considered as a piece of real property, the virgin forest yields no interest ... [if] the ripe timber alone is cut ... the growth of small trees and of those which germinate under the new conditions will be actual added interest" (p. 102).

CONCLUSION

The interweaving of nature, competition, and economic gain painted an attractive picture for Pinchot's audience of professionals and industrialists. In fact, by the end 1899, Pinchot and the (newly renamed) Bureau of Forestry had 123 cooperative arrangements with private landowners and, by 1905, 11 million acres of private forest land were under Bureau management (Steen, 1976, p. 54). With Pinchot as Chief Forester, the Bureau—the best trained and most active foresters in the country—practiced applied forestry as Pinchot had defined it, that is, as he had sketched in the demonstration reports and codified in his *Primer of Forestry* (1899, 1905). It is in the earlier reports, however, that the rhetoric of American forestry was to have its impetus.

Although not expressly intended for a scientific audience, reports such as *Biltmore Forest* and *The Adirondack Spruce* had a lasting effect on the character and *ethos* of applied forestry. For one thing, Pinchot was able in these reports to define forestry as a paradigmatic science, whose claim to validity was based in the ecology of forests and not in a romantic ideal. Like other scientists seeking to establish paradigms for practice, Pinchot had to invent the most efficient rhetorical path; at the time, this involved demonstration in the field and confirmation in writing. When establishing paradigms, Prelli (1989) explains, it is ultimately scientists who must "symbolize rhetorically so that one's favored paradigm is invoked and others are induced to cooperate with it in thought and conduct" (p. 89). The public documents of demonstration forestry show how Pinchot was able to accomplish that goal and secure support for his profession by interweaving typical scientific writing practices with appeals to narrower economic interests and to the broader aims of conservation.

ACKNOWLEDGMENT

The author gratefully acknowledges the assistance of Helen Correll, Margaret FalerSweany, Kathleen Halvorsen, René Johnson, Michael Maıtin, David Martins, Craig Waddell, and Laurie Ann Whitt for their constructive and collegial advice on previous versions and sections of his chapter.

REFERENCES

Clary, D. (1986). *Timber and the forest service*. Lawrence, KS: University Press of Kansas.

Dupree, A. H. (1985). *Science and the federal government*. Baltimore: Johns Hopkins University Press.

Fernow, B. E. (1886). Report of the Chief Forester. In U.S. Commission on Agriculture, *Annual report of the commissioner of agriculture*. (pp.164–167). Washington, DC: U.S. Government Printing Office.

Gross, A. G. (1990). *The rhetoric of science*. Cambridge, MA: Harvard University Press.

Hays, S. P. (1959). *Conservation and the gospel of efficiency: The first American conservation movement, 1890–1920*. Cambridge, MA: Harvard University Press.

Killingsworth, M. J., & Palmer, J. S. (1992). *Ecospeak: Rhetoric and environmental politics in America*. Carbondale, IL: Southern Illinois University Press.

Latour, B. & Woolgar, S. (1979). *Laboratory life: The construction of scientific facts*. London: Sage.

Marsh, G. P. (1965). *Man and nature: Or, physical geography as modified by human action*. Cambridge, MA: Belknap Books of Harvard University Press.

McGeary, M. N. (1960). *Gifford Pinchot: Forester-politician.* Princeton, NJ: Princeton University Press.

Meisner, M. (1995, June). *'Resourcist' language: The symbolic enslavement of nature.* Paper presented at the 3rd Biennial Conference on Communication and Our Environment, Chattanooga, TN.

Miller, C. (1994). *Gifford Pinchot, American conservationist: Two essays by Char Miller.* Lebanon, PA: Grey Towers.

Myerson, G. (1994). *Rhetoric, reason, and society: Rationality as dialogue.* London: Sage.

Oelschlaeger, M. (1991). *The idea of wilderness: From prehistory to the age of ecology.* New Haven, CT: Yale University Press.

Oravec, C. (1984). Conservationism vs. preservationism: The 'public interest' in the Hetch Hetchy controversy. *Quarterly Journal of Speech, 70,* 444–458.

Pinchot, G. (1891). Government forestry abroad. *Publications of the American Economic Association, 6,* 191–238.

Pinchot, G. (1899). *A primer of forestry, part I: The forest* (Bulletin No. 24, U. S. Dept. of Agriculture, Div. of Forestry). Washington, D.C.: U. S. Government Printing Office.

Pinchot, G. (1905). *A primer of forestry, part II: Practical forestry* (Bulletin No. 24, U. S. Dept. of Agriculture, Div. of Forestry). Washington, D.C.: U. S. Government Printing Office.

Pinchot, G. (1970a). *Biltmore forest, the property of Mr. George W. Vanderbilt: An account of its treatment, and the results of the first year's work.* New York: Arno. (Reprinted from original text, 1893, Chicago: Lakeside Press)

Pinchot, G. (1970b). *The Adirondack spruce: A study of the forest in Ne-ha-se-ne Park, with tables of volume and yield and a working-plan for conservative lumbering.* New York: Arno. (Reprinted from original text, 1898, New York: The Critic)

Pinkett, H. (1970). *Gifford Pinchot: Private and public forester.* Urbana, IL: University of Illinois Press.

Prelli, L. J. (1989). *A rhetoric of science: Inventing scientific discourse.* Columbia, SC: University of South Carolina Press.

Steen, H. (1976). *The U.S. Forest Service: A history.* Seattle, WA: University of Washington Press.

Wander, P. (1976). The rhetoric of science. *Western Speech Communication, 40,* 226–235.

8

Rationality Reborn: Historical Roots of the Passive Voice in Scientific Discourse

Dan Ding
Illinois State University

The passive voice is a major stylistic feature of scientific discourse. That is, a scientific writer, when choosing between the active and the passive, is more likely to choose the latter. Closely related to the choice of the passive voice is the avoidance of active constructions with personal subjects—personal active subjects—especially the first persons "I" and "we." In other words, the writer is also more likely to use active constructions with impersonal subjects—impersonal active subjects—if she must use actives at all, because like the passive voice, the impersonal active subjects also emphasize *what* was done instead of *who* did something in scientific experiments. Both of them enable scientific writers to achieve the same rhetorical means (i.e., to "be objective and impersonal") (Herbert, 1965, p. 28). So, the passive voice and impersonal active subjects have received considerable attention from researchers concerned with scientific discourse. Some, like Riley (1991) and Wilkinson (1992), have focused on the rhetorical role of the passive voice. Others like Master (1991), Harmon (1992) and Rodman (1994) have identified either the impersonal active subjects or the passive voice as prevalent features of science writings. However, since most researchers in these studies focused on modern scientific texts (i.e., texts produced in the 1900s) we know little about the passive voice and the impersonal active subjects in early scientific discourse (i.e., scientific discourse as it existed before the 1900s).

So, this need to study early science writings gives rise to at least two questions.

First, were the passive and the active with impersonal subjects prevalent in early scientific discourse? Bazerman (1988) has raised a similar question when discussing the historical analysis of scientific language. He observes that linguists give an impression that "the use of the passive voice and avoidance of the first person" are "timeless expressions of essential character" of scientific discourse (p. 325). Second, if the passive voice and the impersonal active subjects were not regular features in early scientific texts, when did they emerge and why? This chapter is intended to answer these two questions, with emphasis on the second, because I agree with Bazerman who suggests, "We need to understand why [these regular features] emerge ... what the writers accomplish through the use of these features ... why these ... choices have seemed advisable to so many members of the community" (p. 315).

This chapter begins with a review of the passive voice and its related grammatical construct and the impersonal active subjects in modern scientific discourse and examines why those two stylistic features are prevalent in the discourse. Then, it focuses on the early scientific texts—three publications from different periods of time before the 1900s—with an attempt to show when the passive voice and the impersonal subjects started to dominate scientific discourse. Next, it explores the concept of "rationality and reason" and examines how, because of this concept, the passive voice and the impersonal active subjects gradually displaced the active and the first person. Finally, it focuses on the modern scientific discourse community, which embodies the rational values that gave rise to modern scientific discourse.

OBJECTIVITY IN MODERN SCIENTIFIC DISCOURSE

Modern scientific discourse is objective, as Kinneavy (1980) claimed. Although one could challenge this claim by arguing that the ways scientific authors report their research findings "really do not guarantee objectivity" (Vande Kopple, 1992, p. 335), I propose we understand the word "objectivity" in terms of how scientific discourse represents the world. It represents the world in terms of objects, things, and materials. Thus, it appears to be more "object-oriented" (Wilkinson, 1992, p. 322) and "thing-oriented" (Kinneavy, 1980, p. 88) than subject-oriented and human-oriented. Because of this thing-orientation, we can probably expect to see more grammatical subjects that refer to objects, things, and materials than those that refer to humans (researchers), because sentence subjects determine topics of the sentences (Quirk, Greenbaum, Leech, & Svartvik, 1985) and therefore determine whether the sentences are object-oriented or subject-oriented. The passive voice and the impersonal active subjects enable writers to achieve this object-orientation. In this section, I will first attempt to establish that the passive voice and the impersonal active subjects are dominant stylistic features in modern scientific discourse. Then I will explore the relationship between these two stylistic features and "objectivity," relative to the rhetorical setting of scientific writers.

Researchers have provided evidence that the passive voice and the impersonal subjects are indeed prevalent in scientific discourse. Rodman (1994), for example, has studied a 66,500-word corpus of 16 scientific articles and found that 66% are in the passive voice in a corpus of 2,215 transitive verbs. Rodman has also found that of all the active structures, 73% have subjects that refer to the materials, research process, products, and the discourse; only 27% have subjects that refer to humans. In other words, the personal active subjects account for only 9% of all the subjects of the transitive verbs. Harmon (1992), who studied 50 most-cited texts, has found that 53% of the main verbs are in the passive (p. 27), although his corpus of the main verbs includes linking verbs and possibly intransitive verbs. Master (1991), in his study of 10 scientific articles, has found that about 60% of active verbs take either inanimate subjects or abstract subjects (p. 23). So, he observes that "the use of active verbs with instrumental or inanimate subjects is a prevalent phenomenon in English, especially in scientific prose" (p. 15). Indeed, as Vande Kopple's (1994) study has indicated, of all the 780 t-units in 4 scientific articles, only 5.64% have the pronoun "we" as their grammatical subjects and 3.974% have other personal pronouns than "we" (pp. 538–539). Therefore, from these studies we can see that the passive voice and the impersonal subjects appear to dominate scientific discourse.

The articles those researchers studied represent many different areas of science and different periods of time in this century. For example, Vande Kopple's articles cover the fields of physics, chemistry, ecology, and medicine. Rodman's articles were chosen from the fields of mineralogy, chemical engineering, physics, and civil engineering. Harmon's articles represent the fields of biochemistry, analytical chemistry, biophysics, physiology, biology, pharmacology, biometrics, etc., covering an extended period of 60 years from the 1920s to the 1980s. Master's samples cover a period of 10 years from 1975 to 1985. Some articles came from prestigious journals like *Physical Review*, *Journal of Biological Chemistry* and *Journal of the American Chemistry Society*, which are very influential in their respective fields. Some articles themselves exercise significant influence in their fields, as is evidenced by the fact that those articles are very often cited. For example, each of Harmon's articles "received from 5,167 to 187,652 total citations in the 1945–1988 period" (Harmon, 1992, p. 18). Thus, given the variety and influence of those scientific articles, the two stylistic features found dominant in those articles might as well be considered two features dominant in modern scientific articles in a more general sense.

If scientific discourse is objective, I would argue this "objectivity" has a lot to do with the rhetorical setting of scientific authors. Wilkinson (1992) has said that scientists are expected to "disseminate their research for the advancement of science. They do this primarily by writing scientific papers to transmit their new research findings and new ideas to peer scientists" (pp. 319–320). Since scientists "write about organisms, materials, methods, findings, analysis, concepts, etc., not about themselves" and since "they focus on these as objects that they observe, their presentation is said to be objective—that is object-oriented—not subjective" (Wilkinson, p. 322). Wilkinson's idea is that because scientific discourse is intended to transmit one scientist's research findings to other scientists, the discourse should be focused on materials, methods, concepts, etc. Thus, the materials, methods, concepts, etc., are the topic and emphasis in the discourse. Rodman (1994) has also suggested that science writing is about "objects, processes, and phenomena in the physical world" (p. 316). Although in most cases the writer is the scientist who does the research and is thus qualified to make claims like "I placed cells in a thin chamber" (Wilkinson, p. 323), the first person is not appropriate for the scientific writer to transmit the research findings and new ideas because it implies the topic is about "I," the writer, not about the various objects in the research—materials, methods, analysis, concepts, etc. "A paper in the first person is a paper about the writer, not about the research" (Wilkinson, p. 323). Thus, if we consider the rhetorical setting of scientists reporting their findings to other scientists, as Wilkinson argues, the passive voice is particularly appropriate. As Walpole (1979) has pointed out, the main function of the passive is to "retain grammatical focus on a given item of discourse, specifically, on the theme being discussed" (p. 252). So, Killingsworth and Gilbertson (1992) call the passive voice "the thingish style" (pp. 129, 131). They remark that the thingish style is the author's "effort to obtain 'objectivity'" and to indicate that "the events described could occur anywhere" and that "the results are reproducible" (p. 133).

If the passive voice implies "thingish," then impersonal active subjects do too. Master's (1991), Rodman's (1994) and Vande Kopple's (1994) studies have shown that the majority of active subjects in scientific discourse refer to nonhumans. We can thus conclude that scientific writers are as willing to use actives when their subjects refer to nonhumans as they are willing to use the passive voice because, as these three studies suggest, by identifying the nonhuman agents of the actives, the writers are able to make the nonhuman agents the topic of the discourse. So, both the passive voice and the impersonal active subjects maintain the focus on the materials, the concepts, the processes, the analyses, and other nonhuman objects.

By using a large proportion of the passives and impersonal active subjects, scientific writers "make the experimental subject the syntactic subject," which can denote a more general phenomenon, and thus a more universal principle (Riley, 1991, p. 249). On the other hand, the personals don't "universalize the results observed" from experiments; rather, they "present them as idiosyncratic" (p. 248). So, when the personals are not appropriate, scientific writers choose the passive voice and the impersonal active subjects. As a result, we can assume that the frequency of the personals is in inverse ratio to that of the passive and the impersonal active subjects. Depending heavily on those two features, scientific discourse "shifts emphasis from what happened in a specific laboratory to what can be made to happen in any laboratory" and "from the contingent by emphasizing the objective over the subjective, the events of the laboratory over the people who make these events happen" (Gross, 1985, p. 21).

OBJECTIVITY PROBLEM IN SCIENTIFIC DISCOURSE BEFORE THE 1900s

As I pointed out at the beginning of this chapter, there is still much to be learned about the rates of the passive voice and the impersonal active subjects in early scientific discourse. Is it therefore possible that in history the active voice was prevalent with personal subjects, especially "I," in science writings? Bazerman (1988) is probably correct when he argues that the "broad, ahistorical, static identification of features such as the standard five-part structure of the experimental report or the use of the passive voice and avoidance of the first person, are found inaccurate with the slightest amount of historical digging" (p. 315). If the active and the first person were prevalent in early scientific discourse, then the discourse was not objective at all. To find out if the active voice and the first person were indeed prevalent in early science writings, I decided to examine more articles from publications representing different periods of time, hoping for some breadth of representation. I took three publications: *The History and Present State of Electricity Vol. 2* by Priestley, published in 1755 and reprinted in 1966; the *Philosophical Transactions of the Royal Society of London* (Part II, Volume 132 and Part I, Volume 133, published in 1843); and *The Physical Review* (Vol. I published in 1893 and 1894 and Vol. V published in 1897).

To determine the rate of actives in each publication, I compared the frequency of all active transitive verbs to that of all transitive verbs, because only when a verb is transitive will the author face a choice between the passive and the active. As Rodman (1994) asserts, the choice of the passive implies the motivation to suppress the human agent while the use of the active suggests "the willingness or necessity to identify" the human agent (p. 315). So only transitive verbs are significant in this study. To determine the rate of "I," I compared the total number of words in the articles studied to the frequency of "I." Thus, we have the rate of "I" in each publication in terms of one "I" per number of words. To support my findings, I also used another datum: the rate of the active subjects referring to humans. As my above argument suggests, the more personal active subjects there are in a scientific article, the less it is "objective" (i.e., impersonal) because objectivity requires that the topic be on the nonhuman world. This datum is therefore a reliable indication of objectivity of science writings: The rate of the personal active subjects is the frequency of all active subjects referring to humans compared to that of active transitive verbs.

I examined six sections, about 20,064 words, from Priestley's book and found 1,284 transitive verbs (1966, pp. 215–292). In this corpus, 800, about 62.31%, were in the active voice. Obviously, the active is the majority form. Of the 800 active transitive verbs, 412, about 51.50%, took subjects that refer to humans. Because of the high rates of the active and personal active subjects, I also found a high density of the first person. In those six sections, there were 401 cases of "I." When we compare the total number of words to the frequency of "I," we get the rate of "I": 1/50. This means that on an average for every 50 words, there was one case of "I." So, the frequency of "I" is unusually high even if one might argue that some cases of "I" were the author's effort to interpret data or to develop new theories. This high density of "I" prevents readers from focusing on the material world so that constantly they have to focus on the researchers themselves. Thus, every process or action in the experiment was actually manipulated by "I." Let's read a passage:

> After having charged forty-one jars together, each containing about a square foot of coated glass, I let them stand a minute and a half, while I was adjusting some part of the apparatus, in order to make the discharge; when they exploded by the bursting of one of the number. I observed also, that the jar was at a considerable distance from the place when I saw the flash at the wires. ... I am confident that several of mine have burst with a much less charge than they had actually held before. (Priestly, 1966, p. 241)

In this passage, we discern very little of the trait of objectivity because the author "I" did everything, and "I" represents personal and subjective viewpoints. The experiment gives an impression of being personally manipulated. For example, "I let them stand a minute and a half" implies that "I" forced them to stand. Even when "the jar was at a considerable distance from the place," it was "I" that "observed" that it was. But the observation of "I" could have been wrong. When "several of mine have burst with a much less charge," it is only the result of the personal belief of "I," because "*I* am confident" [italics added] that they "have burst with much less charge." In other words, they might not in reality have burst with a much less charge.

I examined one article from the *Philosophical Transactions* Part II, Vol. 132 (Bowerbank, 1843, pp. 215–223), and five articles from Part I, Vol. 133 (Airy, 1843, pp. 45–54; Barry, 1843, p. 33; Belcher, 1843, pp. 55–57; Faraday, 1843, pp. 17–32; Powell, 1843, pp. 35–43), totaling about 27,232 words. There were 1,137 transitive verbs; 601, about 52.86%, were in the active—slightly more than half of all the transitive verbs. In this corpus, the active is still the majority form. Of all the 601 active transitive verbs, 191, about 31.78%, took personal subjects. Compared with Priestley's book, both rates drop in the *Philosophical Transactions*, so we can expect to encounter less frequent "I." Indeed, there were 165 cases of the first person pronoun "I," so the density of "I" is 1/165. That is to say, for every 165 words, there was one case of "I." Clearly, "I" appeared much less frequently in the journal than in Priestley's book.

Personal active subjects disappeared almost entirely at the turn of this century when authors were reporting experiments. With this disappearance came the passive voice and the impersonal active subjects to dominate scientific discourse. I examined six articles from *The Physical Review*, all of which were published in the 1890s: two in 1893 (Marston, pp. 127–138; Snow, pp. 95–116), one in 1894 (Daniel, pp. 241–259), and three in 1897 (Dorsey, pp. 170–181; Foley, pp. 129–151; Millis, pp. 11–20), making up a corpus of about 30,600 words. There were 1,259 transitive verbs; 507, about 40.27%, were in the active form. Obviously in this corpus, then, the active has become the minority form. Of all the 507 active transitive verbs, only 110, about 21.70%, took subjects that refer to humans. I found only 12 cases of "I," so the rate of "I" is only 1/2,550. What is worth noticing is that among the 110 human active subjects there were 34 cases of the first person plural "we" that did not refer to any authors; instead, they all referred to scientists who were reading the articles. For example, in the article "The Study of Distribution of Strains by Polarized Light," Marston (1893) used 31 "we's" to represent scientists in general when he was talking about a result that all qualified scientists could have under the same condition. Take, for example, the following sentence: "By varying the angle of inclination of the polarizing plane and taking a series of observations, *we* [italics added] may obtain the elementary arcs at points as close together as desired" (p. 129). In this passage, "we" does not refer to the author since the author is a single person. Actually, the author uses "he" or "this writer" to refer to himself. So "we" can only refer to scientists in general who are qualified to reproduce the experiment. Here "we" has the same function as the passive voice or the impersonal subjects, which makes a principle or a result universal. The author tried to render his experimental result universal through the use of "we." In addition, he avoided using "I" to refer to himself. My examination of the articles from *The Physical Review* suggests that scientific discourse underwent at least two dramatic changes at the end of the 19th century. First, authors started to depend heavily on the passive voice and impersonal active subjects for reporting experiments; second, the rate of personal subjects, including "I," became very low. There was another change, though a minor one. That is, the authors started to use explicit formal references in the articles, because only in this journal do I find formal references whereby the authors used and acknowledged each other's ideas by identifying the sources of the ideas.

RATIONALITY AND REASON

If we take a second look at the density of the active voice and the personal subjects in the three publications from three different periods, we can see a trace of the development of scientific discourse. That is, the personal features appeared less and less frequently from the 18th century to the 19th century. This decrease in the frequency of actives with personal subjects was not accidental. As I noted above, scientific discourse underwent some major changes at the end of last century before it became modern scientific discourse. The passive voice and the impersonal active subjects started to dominate science writings just before the 1900s. I would argue that this dominance grew out of the concept of "rationality and reason."

In the 17th century, a new idea of science appeared with Bacon, which differed from the medieval ideal and which involved a different conception of truth (Ijsseling, 1976). In his works, especially *The Advancement of Learning*, Bacon (1955a) seeks to survey the methods of inquiry and the present state of knowledge. His ideal model is the inductive practice of a scientist. As Bizzell and Herzberg (1991) have observed, Bacon's "preferred model is the practice of successful scientists, which combines careful observation, experimentation, and classification with an inductive logic that advances learning by revealing natural laws" (p. 623). Here, Bacon "was hardly concerned with gaining control of power over man by means of word, but rather with gaining control of nature through knowledge" (Ijsseling, p. 60). And methods of inquiry about knowledge were important to Bacon. His method is placed in logic, depending on the subject as well as purpose (Zappen, 1975). Bacon's method suggests that "the fortunes of scientific discovery had been overwhelmingly dependent upon a right grasp of methodology" (Bozeman, 1977, p. 7). Bacon himself says that the failure to adapt method to subject reduces learning to "empty and barren generalities" (p. 304). Because of this failure, sciences will be only "husks and shells" (p. 304). Therefore, according to Zappen, Bacon thinks that sciences should "proceed by direct statements and the presentation of data" (p. 245). So, Baconian method is basically inductive. Reid (1969) says that humans, after being taught by Bacon, are at last free from the treadmill of medieval "deductionalism" (p. 145).

Also explicit in Bacon's works is the idea of cooperation, which was "necessitated" by the conception of new science (Jones, 1961, p. 55). This new science was such a huge project that it would require the efforts of generations of scientists to be fruitful. Thus, Bacon (1955a) declares that science is not the work of one man; instead, it "may be done by some person, though not by every one; and may be done by many, though not by any one; and may be done in succession of ages, though not within the hourglass of one man's life" (p. 229). This idea of cooperation is repeated, at least, in *New Organon.* Speaking of the multitude of particulars "facing scientists," Bacon (1955b) suggests that science involves a process "in which the labours and industries of men (especially as regards the collecting of experience) may with the best effect be first distributed and then combined" (pp. 524–525). In these quotations, we detect the budding scientific spirit of cooperation, which, after centuries of evolvement, has blossomed into a major convention of the modern scientific discourse community.

John Locke followed Bacon in searching for truth in the physical world through the method of experimental scientists. Like Bacon, he does not trust deduction, which, according to him, does not generate truth. He thinks that it is possible to understand the outside world "only if we understand the process by which we come to such knowledge" (Bizzell & Herzberg, 1991, p. 697). In the first volume of his *An Essay Concerning Human Understanding*, Locke suggests that our experience is the source of "the materials of reason and knowledge" and that all our knowledge is founded in our experience (Locke, 1959, pp. 121–122). He also suggests that the problem we have in understanding the external world comes from imperfect language. He is very much concerned with presenting the objective world through a clear style. Therefore, Locke endorses the method of seeking knowledge based on induction through "the doctrine of perspicuity and the plain style" (Winterowd, 1968, p. 74). It is clear that during the Age of Reason philosophers like John Locke were influenced by Bacon and were more concerned with seeking truth through reasoning than through sensation. Newton-Smith (1981) has suggested that rationalists tend to examine data "in virtue of how the world is" (pp. 13–14)—in virtue of reality. This rationalistic revolt began in the 17th century with Bacon and gained force in the 18th century during the

Age of Reason, which greatly influenced natural sciences. As Spariosu (1989) has observed, modern science is largely "the offshoot of the Age of Reason" (p. 166). By extension, modern scientific discourse is also the offshoot of the Age of Reason.

When speaking of the rational and pre-rational orientation in the Western World, Spariosu (1989) has argued that in pre-rational mentality, "Power conceives of itself as ... naked and immediate"; authority depends upon "physical strength and cunning intelligence." Also, "competitive values prevail over cooperative values." There is no law to abide by; order is restored through individual authority based on violence. In rational mentality, "power does not present," but "re-present itself," being shared and mediated. Cooperation prevails over violent competition. Written law adjusts social order, and justice becomes a public matter from a private one (pp. 7–8). Based on this argument, we realize that the Baconian notion of seeking knowledge through observation belongs to rational mentality since knowledge is based on presentation of facts, not on personal argumentation, which implies personal enforcement and strength. His idea of cooperation also belongs to rational mentality and also suggests a public image of science. Indeed, Bacon (1955a) once said that science should be more fruitful through "public designation" than "private endeavour" (p. 229). Since the early Renaissance, "most of Western Europe has steadily moved toward a rational mentality" (Spariosu, p. 9) because of the influence of the rationalist rhetoricians' works.

If Bacon started the rationalistic revolt in the 17th century, and the Age of Reason exerted the first influence on natural sciences and scientific writing, then before the end of the 19th century, Darwin's evolutionary theory and the works of H. Spencer with their rationalistic orientation carried on the Age of Reason tradition of rationality and exerted the second and also the major influence on natural sciences and scientific writing. This impact spread to "the majority of natural sciences, such as physics, astronomy, biology, ecology, ethology, and primatology" (Spariosu, p. 165). We can see that the increasing influence of "rationality and reason" paralleled the decrease in the frequency of the active voice with personal subjects in scientific discourse. As a rationalistic concept was becoming increasingly influential, the active voice with personal subjects was becoming decreasingly prevalent.

So, Baconian notion embodying "rationality and reason" became "the prevailing standard" in the mid- and late-19th century science (Zappen, 1990), mainly through the influence of Darwin's and Spencer's works. Let me explain. First, the evolutionary theory shares a major concern with the Baconian notion. Both call for rationality. In his book *The Origin of Species*, Darwin (1936) brings up the notion of natural selection. To be selected by nature, the animate and the plants must struggle, not for power, but for existence. To illustrate this point, Darwin gives examples of both animals and plants. Two canine animals may struggle with each other to get food; a plant in a desert may struggle against the drought for life. This Darwinian notion of "struggle for existence" is obviously more rational, compared with the notion of "struggle for power" which Groos (as cited in Spariosu, 1989) brought up in his book *The Play of Animals*. Second, Darwin's method is based on the rationalist method of seeking knowledge (i.e., based on induction). Campbell (1986), in his research on Darwin's *Origin*, remarks that the strategy, the language, and organization of the *Origin* "appear to recapitulate the inductive ethos of science" and to create "a sense of arriving at Darwin's ideas via the same sorts of observations" (p. 360). We can believe this to be true for, as implied by Campbell, in Darwin's time, the general paradigm of scientific rationality was based on Baconian science: observation and inductive reasoning.

Spencer (1971) carried on this Darwinian theory of natural selection through struggle for existence and further developed the concept of rationality. Echoing Darwin, he claims that evolution is a process of increasing differentiation and integration. Integration refers to mutual interdependence of structurally differentiated parts and coordination of their functions. This notion of integration, of course, implies cooperative values. In *The Principle of Sociology*, he observes that one of the most important principles of natural evolution of civilization is altered from militancy to industrialism (Spencer, 1897). Militancy clearly implies a struggle for power and violence while industrialism suggests cooperation and nonviolence. Here he affirms a society based on voluntary cooperation (Kennedy, 1978, p. 99). So, for Spencer, power in its immediate form of "love of conquest" becomes the "correlative of success in the struggle for existence" (Spariosu, 1989, p. 170).

With Darwin's evolutionary theory and Spencer's works, the concept of "rationality and reason" reached the peak of its development. The Darwinian and Spencerian concepts of evolution, based on Baconian principles, presuppose a rational mentality because the ideas of survival and cooperation are a far cry from violent contest. The notion of evolution, representing the notion of "rationality and reason," was reflected in the sciences as Spariosu (1989) has observed. First of all, the process of conducting research is evolutionary. That is, various theories and results struggle for existence based on disputes and proofs. Tarantino (1991), invoking Cooper, says that scientific information must be recorded and presented precisely, clearly, and objectively "to favor a visual ... representation of facts described" and to help the verification of conclusions by any scientists "qualified to dispute them" (p. 48). The fact that research findings have been reported by one individual scientist does not guarantee that the findings will be accepted by the entire scientific community as valid; they must be reproducible and verified by other scientists before they can be acknowledged as truth. To facilitate this verification, scientists must report the findings in a way that can easily be verified by their peers. That is, scientific information must be described in a way that all qualified scientists can easily test and dispute it. So Tarantino remarks that "only after other specialists in the field have reproduced the phenomenon described and verified the reliability of the information presented can the event reported be accepted as true. On the basis of tested results, new theories and laws can then be generated for use in future predictions and observations" (p. 48).

The process is also cooperative since scientists always generate new theories with reference to old ones. They either use, develop, or refute them. The new theories may also be referenced in the future. According to Barrass (1978), scientific hypotheses and theories must be discarded if found inadequate or if a better explanation is offered "for the accumulated observation on the subject" (p. 25). Here, "the accumulated observation" suggests contributions from many scientists in different fields and different periods of time and thus cooperation among these scientists. So, science makes progress on the basis of disputes and verifications and cooperation among scientists, whereby new theories replace the old and inadequate ones. Thus, Popper (1963) highly praises science, saying that it is perhaps the only human activity in which "errors are systematically criticized and ... corrected" (p. 216). Newton-Smith (1981) calls the scientific method of developing new theories the rational image of science. Modern science appeals to reason but not to a certain authority, to cooperation but not to violent contest. It is a public matter instead of a private one. As part of the evolutionary process of sciences, science writings could not avoid being influenced by "rationality and reason."

INFLUENCE OF RATIONALITY AND REASON ON SCIENTIFIC DISCOURSE

We can discern the effect of "reason and rationality" on scientific discourse in two aspects: a) appearance of journals which, having a public image and carrying cooperative values, gradually replaced books as the major means for scientific publication; b) more importantly, the adoption of the passive voice and the impersonal active subjects in scientific discourse, which has more to do with social consensus and "the intellectual authority of the scientific enterprise" (Gooch, 1975, p. 716) than with personal authority and subjectivity.

The first English scientific journal appeared in 1665 when the Royal Society published its *Philosophical Transactions*. Although, as Bazerman (1988) says, it did not immediately replace the book as the primary means of communicating scientific findings, it represented a major step toward rational mentality. First of all, it was a forum that became more public as contrasted with books and letters. Books represented a private forum; it was a private entity because the forum was owned by the author alone. Letters represented private correspondence, and they implied "a practice of secrecy," in Zuckerman and Merton's (1971, p. 69) words. On the other hand, journals established a public forum and so through publication in a journal, private scientists "became public figures" (Bazerman, p. 134). The significance of the scientific journal is that it encouraged scientific research and aided "flow of information about scientific research both nationally and internationally" (Meadows, 1974, p. 67). Therefore, it "served ... as a public arena where disputes could be hammered out and priority claims established" (Meadows, p. 67). But to hammer out disputes and claim priority, scientists had to disclose their research and works as soon as possible; thus, "The new scientific journal persuaded men of science to replace their attachment to secrecy and limited forms of communication with a willingness to disclose their newly found knowledge" (Zuckerman & Merton, p. 72). So, by publishing their works in journals, scientists transformed their private work into "the most publicly accepted form" (Bazerman, p. 145).

Before the journals appeared, the authors themselves were the judges of their works. They formulated private rules for themselves as these were formed in the pre-rational mentality. But after the journals appeared, an evaluation system slowly but steadily developed. Public judges emerged; they were the "institutionally assigned and ostensibly competent reviewers" (Zuckerman & Merton, 1971, p. 68) and editors. The policies worked out by those people became the public policy, so the editors became an intermediary between the authors and the readers. The authors wrote for the readers and by anticipating the public editorial policy, "authors gained greater control of how their work would be presented" (Bazerman, 1988, p. 133). Thus, their works could better satisfy the readers' needs. From its start, the power of the editorial board never rested in any one individual's hand. Bazerman says that to maintain authority, any one individual editor "cannot be perceived as exercising it, but rather must take a distanced stance on all decisions. ... The invention of editorial boards to handle issues of general policy ... helps the editor maintain authority and trust, but it further allows the journal to establish a corporate identity, representing the field as a whole" (p. 137). From Bazerman's comments on the function of editorial boards, we can understand that the order within a journal system and, by extension, within scientific discourse is maintained by a public policy, not by any one individual, which certainly conforms to rational mentality.

I argued in the previous section that the process of conducting scientific research is evolutionary. By extension, scientific change is also evolutionary. Popper (1963) said, "Continued growth is essential to the rational and empirical character of scientific knowledge," but he emphasized that by "continued growth," he did not mean "the accumulation of observations" (p. 215); instead, he meant "the repeated overthrow of scientific theories and their replacement by better or more satisfactory ones" (p. 215). Popper here seems to suggest that science progresses by advancing new theories which replace old ones. That is, new theories are developed and old ones are refuted and replaced. Scientific change is thus an evolutionary process in which new and valid theories survive while old and invalid ones die. If this evolutionary process involves advancement of new theories and refutation of old ones, then scientists must debate the new theories and laws before they can be accepted as valid. The scientific journal was a very appropriate place for scientists to present and debate new theories. Indeed, from its start, the journal publication was essential to scientific research and "integrated the working scientists into a communications network" (Bazerman, 1988, p. 138). Through the scientific journals, authors presented their own research findings, new theories, the rationale of the theories, etc., so that other scientists could reproduce the research findings, debate the new theories, and either accept them or refute them. Also, through these journals, the authors had access to other scientists' research findings and theories, etc. The journals linked the scientific authors together. Publication in the journal increased the sense of group identity—those who could publish in a journal belong to one group. So, publication in the journal was the first step in becoming a member of a social group; it is an identity-granting social activity (Bazerman, 1988).

Obviously, publication in the journal implies cooperation among the working scientists. As Bazerman (1988) has remarked, when publishing in the journal, "researchers start to recognize the cooperative interlinking of their work" (p. 139). It was then that the reference system occurred as a type of informal recognition of debt (Bazerman, 1988). Meadows (1974) indicates that the formal citation practice began to develop in the late 19th century. This happens to coincide with my findings in the three scientific publications from three different periods; only in the publication from the late 19th century did I find formal citations, although informal references were present in the *Transactions* I studied. The formal citation practice shows that scientists started to be cooperative among themselves. On the other hand, it also reveals that scientists depended upon each other for conducting research and that one scientist's success was dependent on others' works. So citation, even at the beginning, indicated "a close interlinking of the current work within the on-going research and theory which formed a codified network of the literature" (Bazerman, p. 139). Meadows has argued that the citation practice in the latter part of the 19th century was "a sign of the increasing integration of scientific effort" (p. 83). He has also noticed that personal titles in references gradually disappeared before the 20th century and that the use of last names only became the trend. He interprets this trend as an indication of the "increasing impersonality of science" (p. 83). I certainly agree with him, but I would also argue that the trend implied a move away from personal authority to intellectual authority, a move that represented rational mentality. Using personal titles indicates that the authority comes from the title, not necessarily from the scientific research findings. On the other hand, hiding the personal titles forces readers to treat all authors equally and to focus on the research itself.

While the journals gradually replaced the books, the passive voice and impersonal active subjects gradually became dominant in scientific discourse. Just as journals and citation practice conform to rational mentality, the passive voice and the impersonal active subjects also suggest rational mentality because they represent a departure from personal authority and a tendency toward public cooperation and identity with the community. So, they are conducive to helping scientific discourse focus on the objects and phenomena in the real world. In Kies's (1985) words, the passive voice and the impersonal subjects "help maintain the topic of conversation" when the passive sentence's corresponding personal active subjects "would inappropriately place the wrong constituent into thematic position" (p. 304). So, the passive structure and the impersonal subjects suggest that the researchers are emotionally detached from their works (Gooch, 1975) and personally detached as well. On the other hand, personal subjects, especially the use of "I," imply personal authority and personal intrusion in the research. Riley (1991) has claimed that when a scientist uses personal pronouns, the scientist wishes to intrude since the personal pronoun expresses the author's active role in constructing and manipulating the research. However, scientific research "is expected to stand by itself, capable of being repeated, developed, expanded, retested, etc. no matter who performed the original experiments" (Wilkinson, 1992, p. 323). In other words, scientists should affirm the importance of their research over their importance as scientists (p. 323). Personal features should not be present so that the research, the process, the phenomena and the objects "are not a matter of personal interpretation or perspective; they simply may be observed" (Vande Kopple, 1994, p. 554). If, as Schoenfeld (1986) has illustrated, using personal subjects and the active voice has the implication of scientists impulsively forming hypotheses and forcing conditions, then they certainly suggest personal authority rather than intellectual authority of the entire scientific community because the claims they make are based on individuals' manipulation, which implies that the claims are not universally valid in the community.

On the other hand, the passive voice and the impersonal pronouns, which emphasize objects, phenomena, and processes, appeal to intellectual authority, which is shared by the scientific community. An abstention from the personal subjects and the use of the passive voice enhance intellectual authority. The authority goes to the entire community instead of a single person because it implies that the result of a study does not depend on any individual. As long as the research is conducted under the conditions and the principles shared by the community, every qualified scientist should have the same result. The research result reflects general truth, so it can be reproduced by other scientists in the future. And because the result is reproducible by any qualified person, the implication is that it can be integrated by other scientists into their research, thus, again suggesting an interlinking of the current work with future work. This certainly carries with it cooperative values.

MODERN SCIENTIFIC DISCOURSE COMMUNITY

As I noted in the above sections, "rationality and reason" have become increasingly influential in science since the 17th century. To be more specific, the influence is manifest in the cooperation among scientists and the public image of science. Today, we can feel more strongly the influence of "rationality and reason" in the scientific discourse community, not only because it is a community that has two rational images that I discussed in the above sections—cooperation and public identity—but also because it has a third rational image: science is not an infallible force (Randall, 1979). And this rational image is closely related to the cooperative values and the public identity of the modern scientific discourse community.

"The history of science … is a history of … error" (Popper, 1963, p. 216). Here, Popper implies that science makes errors and produces fallible theories. But the fallible theories are often refuted in time. The force that refutes these theories is the intellectual authority of the community, not any individual's personal authority. As I emphasized in the above section, science progresses not just by the simple accumulation of facts; rather, it progresses by advancement of new theory and knowledge and by refutation of old and fallible ones. What makes science rational is "the fact that we can examine [a theory] critically, that is to say, subject it to attempted refutations" (Popper, 1963, p. 221). Therefore, scientists are willing to submit their research results for duplication and verification by the community and to give up a theory that has been refuted by the community as being no longer valid based on new evidence. This willingness implies that the scientists accept the intellectual authority of the entire community as the sole authority.

Science has produced quite a few fallible theories and concepts, but new theories and concepts soon replace them through open presentations, interpretations, and debates. Thus, scientific discourse "takes the form of arguments, aimed at persuading the reader of the correctness of a specific point of view" (Yearley, 1981, p. 410). We usually find most arguments in the discussion section of the scientific article. But the persuasion and the argument are based on presenting facts and taking the audience into consideration. That is, persuasion comes from the cooperation between author and audience. Yearley has pointed out that a scientistic rationality is that "correctness will be associated with inducing from facts" (p. 418) because audience expects scientists' writing to be "rational rather than emotional" and his approach "factual rather than suggestive" (Mitchell, as cited in Baker, 1983, p. 119). Giving factual information allows readers to draw their own conclusions "unhampered by the author's preconclusions" (Enke, 1978, p. 42). Again, we realize that author and reader cooperate with each other in interpreting research findings. As members of the community, both author and audience are on equal plane; the author "has to display himself as a member of the community, with no privileged epistemological access to the realm of objects, and yet he has to show himself as a reporter of the nature of objects in that external realm" (Woolgar, 1981, p. 254).

So scientific authors use the passive voice and the impersonal subjects, enabling readers to focus on the external reality. If audiences are persuaded that a new theory is valid, then the persuasion comes from the external world presented, not from the author. Similarly, if the audiences are persuaded that the theory is inadequate, then the persuasion also comes from the external world presented. Presenting the external world and submitting it to the community for verification suggest that the authors put intellectual authority above their own personal authority; therefore, they are willing to discard their theories if found inadequate by the community. In this sense, the authors' willingness to give up any theory refuted by the community as invalid is made manifest in the passive voice and the impersonal active subjects. Also, the passive voice and the impersonal active subjects send a message that in an experiment all members in a scientific community should obtain the same results under the same conditions. This is why Woolgar (1981) thinks that the idea of membership is implied in scientific discourse because all members would have the same conclusion if given the same conditions. I would argue that the invocation of membership in scientific discourse shows that science appeals to the intelligence of the community as the authority, the authority shared by the entire community. This notion of authority conforms to Spariosu's rational mentality that power is shared and that social consensus is emphasized. Any appeal to individual authority is a departure from rationality. Actually in a scientific discourse community, nobody's subjective observation justifies or establishes the truth of any statement, as Popper (1959) has argued. Popper rejects individual authority not shared by the scientific community. The authority he argues for is one shared by all members of the community. Therefore, Notturno (1985) says that Popper's authority refers to "rational criticism of (one's own theory and of competing theories)" (p. 131). Clearly, for Popper, any scientist's research findings must be able to be challenged and criticized and verified by the community before they can be accepted as truth. Nagel (1979) also argues that progress in science happens not through appealing to any authority, but to a critical methodology: "Ideas which can survive the cross fire of the varied critical commentary that a large number of independently acting minds supply stand a better chance of being sound than conceptions which are alleged to be valid simply because they appear to be self-evident to some individual thinkers" (pp. 13–14). Here, like Popper, Nagel also thinks the authority lies in the method of conjecture and refutation, a method approved by the community. One member's theories and concepts can be established only after being tested by other members of the community. This is what Meyer (1994) calls the "questioning process" of scientific rationality (p. 128).

CLOSING REMARKS

"Rationality and reason" influenced scientific discourse in at least two aspects: the development of journals and the scientific writing style. Because of this influence, the passive voice and its related grammatical construct, impersonal active subject, gradually dominated scientific discourse before the end of the 19th century. The development of the journals paralleled the evolvement of science writing; by the time journals were established as the major forum, the passive voice and the impersonal active subjects had become dominant in scientific discourse. Thus, the development of journals also reveals the importance of "rationality" in establishing the passive voice and the impersonal active subjects as social agency in scientific discourse. It shows that the influence of "rationality and reason" permeated the discourse deeply. Even today we can see the effect of that influence in the modern scientific discourse community. The values the community embodies reflect and are derived from the influence of "rationality and reason," which, of course, constitutes the historical roots of the passive voice and its related grammatical construct, impersonal active subjects. The modern scientific discourse community is established as a public and cooperative identity. So is modern scientific discourse. Thus, the passive voice and the impersonal active subjects are necessitated by the nature of science. Kennedy and Montgomery (1993) have observed that science writing "is determined by the members of a group … the profession"; thus, it implies a "social notion" (p. 3). The passive voice and its related grammatical construct, impersonal active subjects, are just two features that conform to this "social" nature of a scientific community.

ACKNOWLEDGMENT

The author would like to thank the editors of this volume for reading and commenting on the draft of this chapter and Professor James Kalmbach for sharing his ideas on this topic with the author.

REFERENCES

Airy, G. B. (1843). On the laws of individual tides at Southampton and at Ipswich. *Philosophical Transaction of the Royal Society of London, 132,* 45–54.

Bacon, F. (1955a). *Advancement of learning.* In H. G. Dick (Ed.), *Selected writings of Francis Bacon* (pp. 155–392). New York: The Modern Library.

Bacon, F. (1955b). *New organon.* In H. G. Dick (Ed.), *Selected writings of Francis Bacon* (pp. 454–540). New York: The Modern Library.

Baker, C. (1983). Francis Bacon and the technology of style. *The Technical Writing Teacher, 10,* 118–123.

Barrass, R. (1978). *Scientists must write.* London: Chapman and Hall.

Barry, M. (1843). Spermatozoa observed within the mammiferous ovum. *Philosophical Transaction of the Royal Society of London, 132,* 33.

Bazerman, C. (1988). *Shaping written knowledge: The genre and activity of the experimental article in science.* Madison, WI: University of Wisconsin Press.

Belcher, E. (1843). Tide observations at Otaheite, or Tahiti. *Philosophical Transaction of the Royal Society of London, 132,* 55–57.

Bizzell, P., & Herzberg, B. (Eds.). (1991). *The rhetorical tradition: Reading from classical times to the present.* Boston: Redford Books of St. Martin's Press.

Bowerbank, J. S. (1843). On the organic tissues in the bony structure of the corallidae. *Philosophical Transaction of the Royal Society of London, 132,* 215–223.

Bozeman, T. D. (1977). *Protestants in an age of science: The Baconian ideal and antebellum American religious thought.* Chapel Hill, NC: University of North Carolina Press.

Campbell, J. A. (1986). Scientific revolution and the grammar of culture: The case of Darwin's Origin. *The Quarterly Journal of Speech, 72,* 351–376.

Daniel, J. (1894). A study of the polarization upon a thin metal partition in a voltameter. *Physical Review, 1,* 241–259.

Darwin, C. (1936). *The origin of species by means of natural selection or the preservation of favored races in the struggle for life.* New York: The Modern Library.

Dorsey, N. E. (1897). The surface tension of water and of certain oilute aqueous solutions determined by the Method of Ripples. *Physical Review, 5,* 170–181.

Enke, C. C. (1978). Scientific writing: One scientist's perspective. *English Journal, 67,* 40–43.

Faraday, M. (1843). Experimental researchers in electricity: On the electricity evolved by the friction of water and steam against other bodies. *Philosophical Transaction of the Royal Society of London, 132,* 17–32.

Foley, A. L. (1897). Arc spectra. *Physical Review, 5,* 129–151.

Gooch, J. L. (1975). Knowing, communicating, and scientific rhetoric. *BioScience, 25,* 715–717.

Gross, A. G. (1985). The form of experimental paper: A realization of the myth of induction. *Journal of Technical Writing and Communication, 15,* 15–26.

Harmon, J. E. (1992). An analysis of fifty citation superstars from the scientific literature. *Journal of Technical Writing and Communication, 22,* 17–37.

Herbert, A. J. (1965). *The structure of technical English.* London: Longman.

Ijsseling, S. (1976). *Rhetoric and philosophy in conflict: A historical survey.* The Hague: Martinus Nijhoff.

Jones, R. F. (1961). *Ancients and moderns: A study of the rise of the scientific movement in seventeenth-century England.* St. Louis, MO: Washington University Studies.

Kennedy, J. G. (1978). *Herbert Spencer.* Boston: Twayne Publishers.

Kennedy, G. E., & Montgomery, T. T. (1993). *Solving problems through technical and professional writing.* New York: McGraw-Hill.

Kies, D. (1985). Some stylistic features of business and technical writing: The functions of passive voice, nominalization, and agency. *Journal of Technical Writing and Communication, 15,* 299–308.

Killingsworth, M., & Gilbertson, M. K. (1992). *Signs, genres, and communities in technical communication.* Amityville, NY: Baywood.

Kinneavy, J. L. (1980). *A theory of discourse.* New York: W. W. Norton & Company.

Locke, J. (1959). *An essay concerning human understanding* (Vol. I). New York: Dover.

Marston, A. (1893). The study of the distribution of strains by polarized light. *The Physical Review, 1,* 127–138.

Master, P. (1991). Active verbs with inanimate subjects in scientific prose. *English for Specific Purposes, 10,* 15–33.

Meadows, A. J. (1974). *Communication in science.* London: Butterworths.

Meyer, M. (1994). *Rhetoric, language and reason.* University Park, PA: The Pennsylvania State University Press.

Millis, F. F. (1897). An experimental study of induction phenomena in alternating current circuits. *Physical Review, 5,* 11–20.

Nagel, E. (1979). *Teleology revisited.* New York: Columbia University Press.

Newton-Smith, W. H. (1981). *The rationality of science.* Boston: Routledge & Kegan Paul.

Notturno, M. A. (1985). *Objectivity, rationality and the third realm: Justification and the grounds of psychologism—a study of Frege and Popper.* Dordrecht, Holland: Martinus Nijhoff Publishers.

Popper, K. R. (1959). *The logic of scientific discovery.* London: Hutchinson & Co.

Popper, K. R. (1963). *Conjectures and refutations.* London: Routledge & Kegan Paul.

Powell, B. (1843). Observations on certain cases of elliptic polarization of light by reflecvion. *Philosophical Transaction of the Royal Society of London, 132,* 35–43.

Priestley, J. (1966). *The history and present state of electricity with original experiments* (Vol. 2). New York: Johnson Reprint Corporation.

Quirk, R., Greenbaum, S., Leech, G., & Svartvik, J. (1985). *A comprehensive grammar of the English language.* London: Longman.

Randall, A. F. (1979). Scientific writing beyond the textbook. *Science teacher, 46,* 18–21.

Reid, T. (1969). *Essays on the intellectual powers of man.* Cambridge, MA: MIT Press.

Riley, K. (1991). Passive voice and rhetorical role in scientific writing. *Journal of Technical Writing and Communication, 21,* 239–257.

Rodman, L. (1994). The active voice in scientific articles: Frequency and discourse functions. *Journal of Technical Writing and Communication, 24,* 309–331.

Schoenfeld, R. (1986). *The chemist's English.* Weinheim, NY: VCH.

Snow, B. W. (1893). On the infra-red spectra of the alkalies. *Physical Review, 1,* 95–116.

Spariosu, M. (1989). *Dionysus reborn: Play and the aesthetic dimension in modern philosophical and scientific discourse.* Ithaca, NY: Cornell University Press.

Spencer, H. (1897). *The principle of sociology* (Vols 1–3). New York: Appleton.

Spencer, H. (1971). *Structure, function and evolution.* London: Michael Joseph.

Tarantino, M. (1991). English for science and technology: A quest for legitimacy. *English for Specific Purposes, 10,* 47–60.

Vande Kopple, W. J. (1992). Noun phrases and the style of scientific discourse. In S. P. Witt, N. Nakadate, & R. D. Cherry (Eds.), *A rhetoric of doing: Essays on written discourse in honor of James L. Kinneavy* (pp. 328–348). Carbondale, IL: Southern Illinois University Press.

Vande Kopple, W. J. (1994). Some characteristics and functions of grammatical subjects in scientific discourse. *Written Communication, 11,* 534–564.

Walpole, J. R. (1979). Why must the passive voice be damned? *College Composition and Communication, 30,* 251–254.

Wilkinson, A. M. (1992). Jargon and the passive voice: Prescriptions and proscriptions for scientific writing. *Journal of Technical Writing and Communication, 22,* 319–325.

Winterowd, W. R. (1968). *Rhetoric: A synthesis.* New York: Holt, Rinehart and Winston, Inc.

Woolgar, S. (1981). Discovery: logic and sequence in a scientific text (1). In K. D. Knorr, R. Krohn, & R. Whitely (Eds.), *The social process scientific investigation* (pp. 239–268). Dordrecht, Holland: D. Reidel.

Yearley, S. (1981). Textual persuasion: the role of social accounting in the construction of scientific arguments. *Philosophy of the Social Sciences, 11,* 409–435.

Zappen, J. P. (1975). Francis Bacon and the rhetoric of science. *College Composition and Communication, 26,* 244–247.

Zappen, J. P. (1990). Scientific rhetoric in the nineteenth and early twentieth centuries: Herbert Spencer, Thomas Huxley, and John Dewey. In C. Bazerman, & J. Paradis (Eds.), *Textual dynamics of the professions: Historical and contemporary studies of writing in professional communities* (pp. 145–167). Madison, WI: University of Wisconsin Press.

Zuckerman, H., & Merton, R. K. (1971). Patterns of evaluation in science: Institutionalization, structure and functions of the referee system. *Minerva, 9,* 66–100.

part IV

The Language of Science

9

Integrating Multiple Analyses in Historical Studies of Scientific Discourse: *The Philosophical Transactions of the Royal Society of London*, 1675–1975

Dwight Atkinson
Auburn University

> If discourse is nothing less than language itself, and if discourse analysis attempts to admit a broad range of research to the analysis of language, it is by nature interdisciplinary.
> —Tannen (1990, p. 110)

THE PROBLEM

According to Deborah Tannen (1990), discourse analysis is the study of language in context—in all of language's varied and wondrous manifestations as a multifaceted, multifunctional tool of human action and expression.[1] If one is to take this definition seriously, one must not be just willing, but also eager and even aggressive in the search for a rich analytical repertoire—one by which we can realistically hope to capture some of the breathtaking variation and situatedness that human language displays.

In this chapter, I try to show how Tannen's definition of discourse analysis has implications not just for our understanding of language/discourse in general, but for the study of historical scientific discourse in particular. I do so through the application of a combination of discourse-analytic methods from different disciplines to a large corpus of historical scientific texts. Before doing so, however, I briefly develop a rationale for applying Tannen's definition seriously in the study of scientific discourse across history. My discussion is organized around two basic shortcomings of discourse analysis as it is practiced (as opposed to defined) in the present day.

A first major shortcoming of discourse analysis concerns the place that historical studies occupy within it. Tannen urges that as discourse analysts we must seriously consider the full range of contexts in which language naturally occurs, and one of these is obviously the historical context. Just as obviously, however, the historical context has been neglected by and large in discourse analysis (see any standard treatment—e.g., Schriffrin, 1994; Van Dijk, 1985). Instead, emphasis has been overwhelmingly placed on discourse practices occurring in the here-and-now, as if they arose, in some sense, *ahistorically*. But just as surely as all discourse has social and cultural implications, and all discourse is implicated in the exercise of power, so too does all discourse have a history. At best, only distorted understandings are possible when discourse is treated as instantiated merely at the moment of speaking, writing, or signing. By leaving aside the historical dimension—by studying only the tiny present-day tip of what is usually a large historical iceberg—we risk providing discourse descriptions which are seriously incomplete and asymmetrical.

Not only does discourse generally have a past—and one that has meaning for the present—but it is often the historical perspective that allows us most clearly to see how discourse "works." This point has been made more generally regarding the social practice of science (including, but not limited to, scientific discourse) by Latour (1987) in his discussion of "science in the making." According to Latour (p. 4), in studying science as a social practice one can look at either "all-made [i.e., *already-made*] science" or "science in the making." One of the characteristics of already-made science, however, is that the glitches tend to be worked out; a well-done scientific experiment, a well-written scientific report, or a well-established scientific theory present seamless, unitary edifices that resist analysis. Like a well-made movie, products of already-made science have an air of naturalness and inevitability about them. Science in the making, on the other hand, reveals all the messy, contradictory realities that go into constructing a glossy finished product. Just as stepping onto the set of a movie in progress would give a better idea of what goes into making a movie, science in the making provides a truer picture of what goes into constructing the seemingly natural products of science.

The same is true specifically for discourse, especially discourse which is designed, without seeming to, to persuade. By looking at the historical development of scientific discourse, for example, we can see something of how scientists have developed their claim to unique understanding of the natural world—rather than having to take that claim as given. In this way, the study of history—including historical forms of discourse—provides a unique window on the development of dominant social institutions and practices such as science. History is therefore a dimension of context for which discourse analysis should find a central place among its concerns.

A second shortcoming of discourse analysis as presently practiced—and one that Tannen's definition again gives us means to address—is a methodological one. While there are certainly many methods of discourse analysis, since the field is broad and heterogeneous, there is a strong tendency toward the exclusive use of one or a small number of methodologies in particular areas of study. This reality has at least two implications: a) that such exclusive uses of one or a few methodologies will tend to "lock" analysts into one or a few ways of looking at the object of study; and b) that inasmuch as a single methodology becomes the special property of an area or a field, and thereby undergoes independent development in that field, its usefulness to discourse analysts in other fields, and the relevance of its results to other fields, will diminish accordingly.

Although in a somewhat different context, the linguist William Labov (1971) has suggested a partial solution to these problems in the *combining of diverse research methodologies*.[2] According to Labov:

> Good linguistic work will draw upon several types of data and methods in approaching a given problem. Since different methods will have different sources of error, even a partial convergence in the results will lead to a higher degree of confidence than we can give to work with one kind of data and one way of dealing with it. (pp. 478–479)

Looking at this quotation in the context of Tannen's definition of discourse analysis, we could say that the former puts the latter in a new light: That discourse as a complex and multifaceted human endeavor can best be studied by applying multiple methodological approaches from differing disciplines in its analysis.

THE RESEARCH

For the past half-dozen years I have been applying a multidisciplinary, multimethods approach in the study of written scientific and medical discourse across history (Atkinson, 1992; Atkinson, 1993; Atkinson, 1996; Atkinson, in press). In this work, I bring methods of discourse analysis from two diverse fields—linguistics and rhetoric/composition—to bear in an attempt to gain a fuller and better grounded perspective on the historically influential texts I study.

The texts I have been investigating most recently are a large, cross-historical corpus of scientific research articles from the *Philosophical Transactions of the Royal Society of London* (hereafter *PTRS*). These texts are sampled at seven 50-year intervals from 1675 to 1975, substantially the full existence of both the Royal Society and its journal. This corpus was designed as a representative sampling of articles from the

PTRS across the full historical period studied (for details of corpus construction see Biber, Finegan, & Atkinson, 1994; Atkinson, in press).[3]

According to the eminent 19th-century biologist Thomas Henry Huxley:

> If all the books in the world, except the "Philosophical Transactions," were destroyed, it is safe to say that the foundations of physical science would remain unshaken, and that the vast intellectual progress of the last two centuries would be largely, though incompletely, recorded. Nor have any signs of halting or of decrepitude manifested themselves in our own times. (1866/1968, p. 23)

On the basis of expert opinions such as this one, and the well-documented influence of the Royal Society and the *PTRS* on modern science over the last three centuries (e.g., Atkinson, 1993, chap. 3, note 2 and Appendix 1; Bazerman, 1988, chap. 3), it can be claimed that the study of a representative corpus of articles from the *PTRS* is a study of written scientific discourse in a more general sense, as it has developed from the beginnings of modern science to the present day.[4]

THE METHODS

The first type of discourse analysis used in this work is what I call rhetorical text analysis, or *rhetorical analysis,* for short. This is the form of discourse analysis overwhelmingly adopted by composition, rhetoric, and technical writing researchers in the study of scientific discourse (e.g., Bazerman, 1988; Berkenkotter & Huckin, 1995; Myers, 1990). Although rhetorical analysis is not easy to characterize as a single, unified discourse-analytic method, it does have some important identifying features.

First, rhetorical analysis is *eclectic*. That is, although it has its roots in rhetoric/composition, it borrows concepts and techniques from a broad range of fields. The only requirement for such borrowing is that the concepts or techniques thus obtained have potential to shed new light on the object of study. In my own version of rhetorical analysis some of the concepts originate in linguistic and applied linguistic research (e.g., Atkinson, 1991; Ferguson, 1994; Swales, 1990) while others derive from the work of composition/rhetoric researchers, literary theorists, philosophers, and historians and sociologists of science.

Second, rhetorical analysis is *contextual* in that it depends substantially on knowledge of the workings of the discourse communities in which the texts under study are or historically were situated. Where the discourse community is one presently in existence, the researcher therefore frequently engages in some level of participant observation (Spradley, 1980); recent studies of situated literate practices in the sciences and beyond (e.g., Berkenkotter & Huckin, 1995; Brown & Herndl, 1986; Freed & Broadhead, 1987; Heath, 1983; Knorr-Cetina, 1981; Latour & Woolgar, 1986; Myers, 1990) provide strong exemplars for this approach. Where, on the other hand, the discourse community is a historical one—as in the present study—researchers must immerse themselves deeply in the available historical sources and secondary literature. It is only in this way that one can hope to gain the articulated understanding necessary to perform an analysis of situated linguistic and rhetorical practice. As one well-known historical discourse researcher has put it, it is impossible to "see what a text was doing without looking at the worlds in which these texts served as significant activity" (Bazerman, 1988, p. 4).

Third, rhetorical analysis is *interpretive,* assuming that the researcher can "read off" important aspects of the surrounding context from the text itself. To employ a scientific analogy, in the same way that ecologists assume that "an animal's wings, gills, snout or hands describe an animal's environment" (Michaels & Carello, 1981, as cited in Rogoff, 1990, p. 30), so the situated written word is assumed to reflect its complex environments of production and consumption. This assumption is made explicit in recent text-focused rhetorical research (e.g., Bazerman, 1994, p. 93; MacDonald, 1994, p. 7). As used in the present analysis, however, it must be noted that the analytical reading of texts as deeply embedded in their sociohistorical contexts (i.e., point 2, above) and the "reading off" of such contexts from the texts themselves are mutually supporting, dialectically related activities.

Fourth, rhetorical analysis is a *bottom-up,* or inductive, type of analysis, in which results are "emergent." This is in line with other qualitative approaches to empirical research based on hermeneutic and phenomenological perspectives.

Fifth and finally, rhetorical analysis operates, theoretically speaking, at the level of *genre.* In my understanding of the concept, genres represent conventionalized whole-message forms which correspond to what Goffman (1972) called "frames"—or primary frameworks for interpreting sense experience. Genres represent such primary frameworks in the rhetorical-linguistic domain. More explicitly, genres constitute the external counterparts of intersubjective models or typifications for interpreting discourse (Miller, 1984), and it is only by applying such models that we can narrow down—and produce approximate agreement upon—the apparently infinite number of possible interpretations we might otherwise give any particular text. Looked at from a somewhat different angle, genres are used to "do stuff"—to perform, in the words of Miller, various kinds of social action.

The second method of discourse analysis which I apply in the historical study of scientific writing is multidimensional register analysis, or *MD analysis* (Biber, 1988). While rhetorical analysis takes genre as its organizing notion, MD analysis is based on the sociolinguistic concept of *register* (see Biber & Finegan, 1994). A register, according to Halliday (1988), is "a cluster of associated [linguistic] features having a greater-than-random ... tendency to co-occur" (p. 162). *Co-occurrence of linguistic features* (Ervin-Tripp, 1972) is the key concept in register theory—it is what gives a particular register of text (e.g., scientific English) its characteristic "texture" or style, allowing "any speaker of [the language of the text] for whom it falls within the domain of experience to know [that register] when he sees it or hears it" (Halliday, p. 162). Looked at more globally, registers are varieties of language that occur in situations of varying contextual specificity, and on which these situations and contexts have a formative influence. But while I make a theoretical distinction between the basically linguistic level at which register operates, and the basically rhetorical level at which genre operates, in natural textual practice they occur together, with register being for all practical purposes an indispensable part of genre since it plays a central role in solving the rhetorical problems that genres are constituted to address.

In addition to its theoretical dependence on the concept of register, there are four major identifying characteristics of MD analysis. First, MD analysis relies on the grammatical/lexical tagging and counting capabilities of computers to analyze large, systematically constructed text corpora. Although such "automatic" analysis by no means eliminates human involvement, it has the distinct advantage of allowing much larger numbers of texts to be analyzed than has hitherto been possible (Stubbs, 1996).[5]

Second, MD analysis employs the multivariate statistical technique known as factor analysis (see Biber, 1988, chs. 4 and 5). According to Lauer and Asher (1988, p. 247), factor analysis "extracts an underlying, reduced set of major conceptual variables from a set of real, observed, and interrelated variables." In this case, the real, observed variables are about 60 syntactic and lexical features that have been found in past linguistic studies to distinguish various kinds of texts, and which Biber (1988) analyzed for their frequencies of occurrence in a large standard corpus of present-day spoken and written English texts. The underlying reduced variables are five sets of linguistic features—derived by Biber from the above-mentioned 60 features via factor analysis—that co-occur with a high degree of frequency across these texts. Using these empirically established factors as baseline data, then, I study cross-time register variation in the scientific research article in regard to these same feature sets.

Third, the analytical tools just mentioned allow for the precise quantification of linguistic variation across time within particular types of discourse. Such precision is usefully complementary to primarily qualitative, interpretive methods of discourse analysis such as rhetorical analysis.

Fourth—and very important—the precise numerical descriptions of texts and text groups that are its direct output do not remain disconnected from the texts themselves in my use of MD analysis. In other words, these numbers—and the sets of linguistic features they represent—are used as tools by which one can look back at and further understand what is going on in the texts being studied. This final, crucial step in MD analysis is exemplified in the findings reported below, and further technical information on this necessarily complex method of analysis is given in note 14.

THE FINDINGS[6]

Rhetorical analysis: At the rhetorical level of analysis, a first finding was that research articles in the 17th and 18th centuries exhibited a strong "authorial persona." Authors in the 17th-century *PTRS* presented themselves as actively involved in the events described in their texts, freely referring to themselves in the first person and describing their own thought processes and feelings. The following passage from the middle of an "experimental discourse," published anonymously but later attributed to Robert Boyle, exemplifies this "author-centered prose":

> I doubt not but what I have related and hinted has given you a Curiosity to know somewhat further of this Mercury: And I confess, that if there be any truth in what some of the most approved *Spagyrists* have deliver'd about a Solvent of Gold that seems of kin, and perhaps is not much Nobler than one that I had; it seems allowable to expect, that even ours should be of more than ordinary use, both in *Physick* and *Alchymy.* But the misfortune I had to have lost a considerable quantity of it, being afterward increased by the almost sudden death of the only Operator I trusted in the making of it; I was altogether discouraged from repeating such a troublesom preparation, especially being diverted by Business, Removes, Sickness, and more pleasing studies. (Anonymous, 1675, p. 515)

First-person personal pronouns and active-voice verbs are common in this passage, giving it much of its author-centered flavor. So also is language portraying the affective states and psychological processes of the author (e.g., "doubt," "expect," "discouraged," and "troublesom"). Other rhetorical characteristics linked to the "author-centered" approach in the early *PTRS* include: a) *witnessing*—or the naming of (typically important) persons who attended the scientific event being reported; b) *expressions of modesty and humility*, by which researchers showed diffidence and caution in the interpretation of natural phenonema (cf., Shapin, 1984); c) *a tendency toward miscellaneity*, such that apparent digressions are frequent, and some articles are simply collections of unconnected observations; and d) *elaborate politeness*, as when articles open with encomia to the Royal Society membership, or when fellow researchers are referred to as "ingenious" or "intelligent."

Following its first appearance in the 17th century, the "author-centered" approach appears with nearly equal strength in the 18th, although the highly mathematical nature of some articles in the 1725 volume—when Newton presided over the Royal Society—seems to temporarily override it. In 1775, though, most writers, including well-known scientists like Joseph Priestley, can be found operating fully within the author-centered discourse. There are indications, however, that a different, essentially *non-author-centered* rhetoric was also under development at this time; it appears particularly where the methodology and conduct of experiments are being described:

> A leaf taken from a growing bean was put into the cold mixture, and frozen. … Another fresh leaf was taken and bent in the middle upon itself; a small shallow leaden vessel was put upon the top of the cold mixture, and the two leaves put upon its bottom; but one-half of each leaf was not allowed to touch the vessel by the bend; the cold mixture was between 17° and 15°, and the atmosphere at 22°. (Hunter, 1775, p. 451)

The agentless passive as the nearly exclusive finite verb-form here seems distinctly modern and "object-centered"; other such descriptions in this article, however, show that the writer was not averse to a more author-centered style of reporting when it suited him.

In 1825, the next period sampled, the author-centered approach again appears in force. By 1875 however, much author-centered prose is being used for quite specific rhetorical purposes, or is confined to specific textual locations in reports. The new norm is for discourse which focuses on others' research findings, or which treats natural phenomena in a basically "object-centered" way:

> *Anguis fragilis and Lacerta viridis.*—The description of these two forms may be most conveniently taken together. ... The area of tooth-development exists in them as a sharply defined region, bounded on its inner side by connective tissue, just as in the newt and the frog; but although it is not restricted by extraneous causes, such as the antagonism of the upper and lower jaws, it nevertheless is not widely extended as in the newt. (Tomes, 1875, p. 292)

Also appearing in such descriptions, however, is author-centered discourse functioning in a specialized way: to index the author's uncertainty—or his/her unique position—regarding some aspect of the text's content (cf., Tarone, Dwyer, Gillette, & Icke, 1981). This is seen late in the paragraph following the above passage:

> The enamel-germ appears to be given off from that of the preceding tooth-sac (fig. 19); at least a process is very often discoverable at the side of this latter, although the connexion with the oral epithelium is not lost and appears to be tolerably direct (see figs. 18 & 16): I am inclined to think that the enamel-germs do not arise from the oral epithelium quite *de novo*. (Tomes, 1875, p. 292)

A second specialized function of the author-centered style in 1875 is to begin articles or specific sections within them. Just as frequent in this period, however, are articles in which the author is either "distanced" at article beginning through the use of such devices as third-person self-reference (e.g., "the author"), or appears to be left out altogether.

Finally, articles from 1925 and 1975 show continued development of distanced and "effaced" author norms, as in article beginnings:

> Early in 1923 it was shown that a tungsten filament heated to 1200° K ... in saturated caesium vapour converts all caesium atoms which strike it into caesium ions. Thus when the filament is surrounded by a negatively charged cylinder a positive ion current flows from the filament, ... if this is sufficient to overcome the space charge effect of the positive ions. (Langmuir & Kingdon, 1925, p. 61)

That this is more than just object-centered discourse—that direct indexes of the author are in fact being *avoided*—is made clear by the fact that "it was shown" in the first line refers to earlier work by these same authors. In other cases, authors refer to themselves exclusively as "the author," extending this term even into post-article acknowledgments. However, despite the fact that articles have patently become more object-centered and "impersonal" in the 20th century, it is not true that author-centered writing has completely vanished from the *PTRS*. The purely theoretical paper written by physicists or mathematicians, for example, is one article type in which the author is given comparatively direct expression. An even smaller number of articles, all of them biological, retain a generalized, if muted, author-centered rhetoric. By far the larger number of articles in 1925 and 1975, however, include few if any direct indexes of the author.

A second finding at the rhetorical level of analysis was that the genre forms in which research articles usually appeared in the *PTRS* changed radically over the 300 years studied. Thus, while a large number of apparent genre names occurred in article titles in the 17th and 18th centuries,[7] the "polite letter" was by far the most common discrete form in which research was actually reported.

Letters represent approximately 50 percent of all articles published in two of the first three *PTRS* volumes examined here (i.e., for 1675 and 1775), and 33 percent in the 1725 volume. These letters typically begin with the salutation "Sir/s," have introductions and sometimes closings honoring their addressees, and tend to be short and somewhat miscellaneously organized. The following—addressed to the *PTRS*'s founding editor, Henry Oldenburg—is one of the more elaborate instances:

> Honoured Sir,
> Having long understood the freedom of address, you have allowed to any candid and unprejudic'd persons, whose general good wishes to the Common-wealth of Learning make 'em ambitious to impart something, whereby they may promote the Empire of Man above other Creatures; I flatter'd my self, it would be no unpardonable presumption, to communicate some of my thoughts unto you, on such a subject, as that the usefulness of the matter may keep me from seeming altogether impertinent, and the smalness of my performance, argue the greatness of my desires to serve you. This I resolved to do by way of Animadversions on the two last *Philosophical Transactions*; and though I performed nothing more, I hop'd at least, my Observations may do them the kindness, as shades intermixt with light, to render them more conspicuous. In hopes of this, I shall draw up my thoughts under the general Title of *Some Philosophical Observations on some passages in the two last Months Transactions*, viz. for *Decemb.* and *January*. (Anonymous, 1675, p. 533)

Honoring the Royal Society and humbling the author, this opening shows the pronounced politeness identified previously as associated with an author-centered rhetoric. A second characteristic mentioned previously and common to letters from this period—a tendency toward miscellaneity or digression—occurs when the author comments on a variety of subjects featured in two earlier issues of the journal. At least one-third of letters in 1675, most on "natural history" topics, show similar evidence of miscellaneity. Finally, letters in this period are typically quite short, with most no longer than three printed pages.[8]

As noted above, the next period, 1725, shows a significant drop in relative numbers of letters published, a phenomenon which can partly be attributed to Newton's domination of the Royal Society at this time.[9] Newtonian science does not appear to have favored the letter form, given as it was to comparative brevity and miscellaneity. Of the nine articles in the 1725 volume which *are* letters, few exhibit the honoring introductions and miscellaneity which were characteristic in 1675.

By 1775, the next period, some of the characteristics noted for 1675 letters have reappeared. Epistolary articles typically begin "Sir" or "Dear Sir," and frequently include polite/honoring introductions, although both their length and elaborateness are much reduced. Digressions are also again in evidence, although now occurring mostly near the ends of papers. New uses of the epistolary form also appear: A few articles have short "cover letters" accompanying longer, typographically separated articles, and others are composed simply of multiple letters.

The period 1825 is the last one in which letters appeared in the *PTRS* volumes analyzed here. While they still represent roughly one-third of all articles, letters in this period show important differences. First, letter beginnings are only sometimes honoring—they are just as likely to "get straight to the point":

> DEAR SIR,
> It is more than two years since, in a conversation I had with you on subjects connected with magnetism, you enquired what effect I thought might result from giving to an iron ball a rapid rotation? The subject however dropped, and it did not occur to me again, till in some speculative views in which I was lately engaged, as to the cause of the rotation of the earth's magnetic poles, the apparent irregularity of the terrestrial direct powers, &c. I was led to consider that, probably, rotation might have a certain influence. (Barlow, 1825, p. 117)

While this letter retains a personal tone, the emphasis here, as in other article openings in this period, is most directly on the topic at hand. A second characteristic of letters in 1825 (as with articles generally) is their greater length, although short letters can still be found. Finally, less than half of the letters in this period are on natural history subjects.

A third finding at the rhetorical level of analysis concerns the public stance authors took toward their discourse communities. Articles in the 17th and 18th centuries are "dialogic"—they frequently address personally known parties and tend to foreground overt cooperative purposes. Thus, articles in 1675 frequently present or answer lists of "queries" sent out by Royal Society members seeking information about various aspects of the natural world.

A second indicator of discourse community dialogue and cooperation in 1675 is the many articles written in direct response to other *PTRS* reports. This phenomenon is evident in titles and openings, e.g., *An extract of a Letter, written by a Friend to the Publisher ... relating to the Contents of the Tract next foregoing* [sic]. Also apparent in this title are two additional markers of discourse community orientation: author anonymity, a characteristic attributed by Kronick (1988) to an ideal of communistic, self-abnegating science; and the close personal relationships that then obtained in the Royal Society. A third point relating to the discourse community is that around 20 percent of articles in both this and the following period are in Latin, still at that time the international language of learned communication.

In the next period sampled, 1725, the dialogic nature of *PTRS* articles is again apparent; the "dialogue" however is no longer necessarily a cooperative one. Instead, authors relate their work to that of others in a more polemical manner than previously. This "oppositional" approach is sometimes immediately indicated, for example, in titles and openings:

> *An Enquiry into a Discovery, said to have been made by Signor Valsalva of Bologna, of an Excretory Duct from the Glandula Renalis to the Epididymis.*
> The late learned Signor Valsalva having some Time ago discover'd a Vessel, which he took to be an Excretory Duct ... ; I lately procur'd a human Body, in order to search for it, which I did with all the Diligence I was capable of. I was not so happy as to discover any Duct of this Kind. (Ranby, 1725, p. 270)

A second marker of oppositional discourse in 1725 is the manner in which writers anticipate disagreement, using such phrases as "But here it may be objected that" to introduce a potential counter-argument, which is then defeated. The use of "manners in dispute" (Shapin, 1984)—marked in the above passage, for example, by the attribution "*learned* Signor Valsalva"—is another feature of oppositional discourse in this period, in which the Royal Society was guided by an agonistic Newtonian agenda (Heilbron, 1983).

In the next period, 1775, many articles have recovered the cooperative dialogic tone described previously. In addition, authors commonly take two other stances vis à vis their readers, either portraying themselves as simply presenting information to the Royal Society in general, or taking a more-or-less "self-absorbed" approach. The former may represent a neo-Baconian stance of seeking to add to the general store of natural knowledge, while at the same time avoiding the more personalized "dialogic" orientation; in the latter, authors seem to pursue their work primarily for personal reasons—a motivation conflicting directly with the very idea of scientific societies and scientific journalism.[10]

By the next period, 1825, there is evidence of a revolutionary new dynamic forming in the discourse community. Research problem-statements and statements that topics are important or interesting are common, and there are occasional (albeit tentative) calls for further research. Such features may indicate that a core of aficionados existed for at least some scientific research problems and that authors expected their own work to be followed up by others. A few articles even have primitive literature review sections mentioning investigations in similar areas. A second revolutionary feature of texts in this period is a concern for proper scientific methods and instrumentation; far more than before, texts of all kinds emphasize the important place of methods and instruments in scientific research. Several articles even focus *totally* on methods or instrument descriptions, and two have what appear to be incipient "methods sections."

By the next period, 1875, there has been extreme development in the direction of research becoming contextualized in research communities working on similar problems. Literature reviews are relatively common and typically exhaustive. Problem-statements and attempts to link projects to more general concerns are also prominent:

> The investigations which form the subject of this memoir have occupied our attention for a considerable time. ... They have been made collaterally with a series of experiments carried on by a Committee appointed by the Secretary of State for War, with the view ... of determining the most suitable description of powder for use in heavy ordnance, which is still increasing in size; indeed our main object has been to endeavour to throw additional light on the intricate and difficult subject under investigation by that committee.
>
> There are perhaps few questions upon which ... such discordant opinions have been entertained as upon the phenomena and results which attend the combustion of gunpowder. (Noble & Abel, 1875, pp. 9–10)

A ten-page literature review, summarizing 21 earlier investigations, directly follows this opening. Likewise, descriptions of scientific methods and instruments are given even more prominence in this period. In fact, the most notable aspect of *PTRS* articles in 1875 is their apparent exhaustiveness in *all* aspects of scientific research. Thus, the article just quoted is 104 pages long and has 26 discrete sections, lettered from A to Z. In addition, it includes an 18-page appendix giving narrative accounts of 86 experiments, as well as 24 tables and many charts. While this article marks an extreme even in 1875, it clearly represents the then-dominant approach to reporting research.[11]

In the next period, 1925, there has been discourse community development and innovation in several areas. First, articles in general have become substantially more theoretical. Second, and relatedly, articles give somewhat less attention to the close delineation of research procedures and methods; this is particularly true for experimental reports but holds across all types of articles. Third, rhetorical conventions for embedding research in contextual "webs of relevance" (Knorr-Cetina, 1981) have continued to evolve. First noted in 1825, devices such as importance-statements which locate work within research communities have developed significantly, to the extent that they appear across all kinds of articles. One common sequence of claims occurring in article beginnings resembles that identified by Swales (1990) for the present-day research article: *statement of topic's importance/centrality → statement of "gap" in research on topic → "gap-filling" introduction of author's own research.* Although the rhetorical system of locating one's work within a specific research tradition is still actively under development in 1925, its relative conventionalization marks its rapidly growing importance.

In the last period covered in the present analysis—1975—nearly all developments described for 1925 have progressed further. All types of articles—but especially in physics/mathematics—now show a greater concern for theoretical than for empirical matters.[12] There is likewise a further downplaying of research methods and procedures; even in articles over 50 pages long, methods sections are generally limited to no more than two pages.[13] Finally, the use of article-beginning "rhetorical moves" (see above) has also undergone further development. While realized variably, the three-part move sequence described above is adhered to substantially in most articles that are not purely theoretical.

MD analysis:[14] A first finding at the linguistic level of analysis concerns a dimension of discourse labeled "involved versus informational production." "Involved production" describes language of the sort typically produced "online" in casual and interactive contexts, such as face-to-face conversation among equals, whereas "informational production" signifies informationally dense language usually produced in conditions allowing deliberate planning, as often found in contemporary professional writing. As seen in Figure 9.1, *PTRS* texts start in 1675 from a norm marked by low levels of "involved production," then proceed linearly and without exception to one marked by very high levels of "informational production." In fact, the texts from 1975 have a score rivaling the most extreme scores found by Biber (1988) among the 23 genres of modern texts he studied. Although this shift appears gradual, if unswerving,

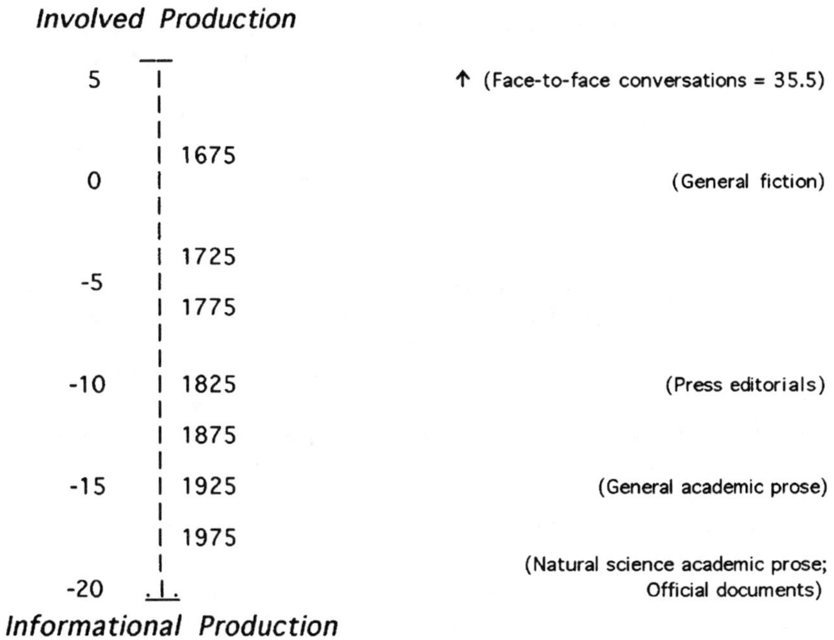

FIGURE 9.1. Dimension 1 scores for *PTRS* research articles by period (parenthetical descriptors indicate genres with comparable factor scores from Biber, 1988). ©1996 Cambridge University Press. Reprinted with the permission of Cambridge University Press.

when viewed from a period-to-period perspective, it represents profound change across the 300 years studied.

The pronounced evolution shown by *PTRS* articles on Dimension 1 can be illustrated by comparing text samples from both ends of the 300-year period under study. In these passages, italics mark "involved" features while bold type marks "informational" features:

> 1675:
> **After** this **manner** *I* used to try the **Experiment:** *For I* have try'd *it* often; sometimes to *observe* the **circumstances of** *it,* sometimes in order to further **Experiments,** and sometimes to show *it* **to** others, and **in** all *my* **tryals** the **success** *was* the same. But whereas **Mr. Line** *thinks, [Ø-THAT] I* tryed *it* **in** a **cloudy day,** and placed the **Prism at** a **distance from** the **hole of** the **window;** the **experiment** will not succeed well if the **day** *be* not clear, and the **prism** placed close **to** the **hole,** or so near at least, that all the **Sun's light** that *comes* **from** the **hole** *may* pass **through** the **Prism** also, so as to appear **in** a **round form** if intercepted **by** a **paper** immediately after *it* has past the **Prism.**
>
> When **Mr. Line** has tryed *this, I could wish, [Ø-THAT]* he would proceed a little further to try *that* which *I* call'd the **Experimentum Crucis,** *seeing [Ø-THAT]* (if *I misremember* not) he denies *that* as well as the other. (Newton, 1675, pp. 500–501)
>
> 1975:
> To account **for** the **maximum slope conductance-voltage curve in** the **nodal membrane in terms of Boltzmann's law** an **undetermined number of charged particles** transfering [*sic*] 3.5 **electronic charges per channel** would have to move **from** a **blocking to** an **open position through** the **entire electric field within** the **membrane** (Fishman, Khodorov & Volkenshtein 1972). *Since* the **effective charge per particle** calculated **from** the **steady state distribution (item** 3 **in table** 2) *is* 1.65 **electron charge** the **ratio**
>
> $$3.5/1.65 = 2.1$$
>
> *represents* the **number of charged particles** which would have to be displaced **from** a **blocking to** an **open position** to activate each **sodium channel.** The **coordination number of** 2 derived **from** the **properties of** the **asymmetrical displacement currents** *is* therefore **in good agreement with** the **results from kinetic data on activation of** the **sodium system** where the **variable** m was raised **to** the **power of two** (Frankenhaeuser 1960). The **data for** the **squid axon** *are* also **in agreement** but **for** a **coordination number of** 3. (Nonner, et al., 1975, pp. 491–492)

These texts well exemplify major changes in the language of the research article that took place between 1675 and 1975. Early texts, like the 1675 passage from an article by Newton, were comparatively "verbal" (Wells, 1960) in that they contained a broad range of verbs operating with their full semantic potentials and occurring frequently in the active voice. Other "involved" features—co-occurring first-person personal pronouns, private (i.e., emotion and cognition) verbs, and *that*-deletions—also indicate that this passage has a somewhat interactive and possibly less planned character. At the same time, however, and not surprisingly given its overall informational purpose, the 1675 passage contains a moderate number of nouns and prepositions. The 1975 passage, on the other hand, is overwhelmingly "nominal" and "informational"—in virtually the same total number of words, it has nearly twice as many nouns as the 1675 passage. There are likewise a profusion of other noun phrase elements such as attributive adjectives and prepositions, making for a large number of "markedly long noun phrases" (Vande Kopple, 1994).[15] Finite verbs, on the other hand, are few and far between in the 1975 passage, and those that do occur exhibit, for the most part, greatly reduced semantic functionality by simply relating large nominal complexes to one another (cf., Halliday, 1988). In sum, register differences between these two text samples relate to the extremely "informational"/nominal nature of the modern-day passage versus the relative balance of "involved" and "informational" features in that from 1675.

A second finding at the linguistic level of analysis concerns a discourse dimension differentiating texts in terms of "narrative versus non-narrative concerns." Present-day genres with high positive scores on this dimension include romantic, mystery, and science fiction, while professional letters, academic prose, and official documents are among the genres with the highest negative scores (Biber, 1988). A single set of linguistic features underlies Dimension 2—past tense, perfect aspect, and public verbs (e.g., *say*, *mention*) combine with third-person personal pronouns to mark a text as "narrative." Figure 9.2 presents a schematic view of change on Dimension 2 across the 300-year period of study.

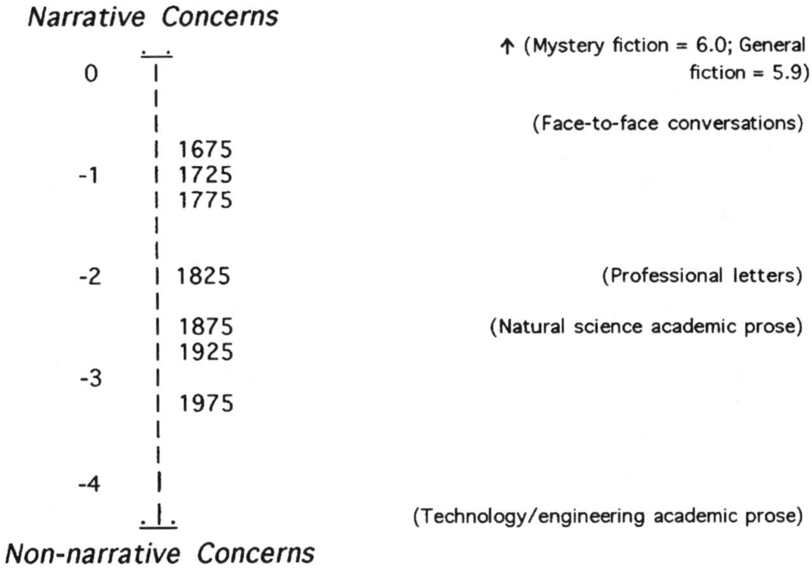

FIGURE 9.2. Dimension 2 scores for *PTRS* research articles by period. © 1996 Cambridge University Press. Reprinted with the permission of Cambridge University Press.

Dimension 2 shows a pattern of evolution which is similar to that of Dimension 1 in terms of its steady progression in a single direction. Scientific research writing in the *PTRS* starts at a low level of "non-narrativity" and becoming progressively more "non-narrative" over time, until it approaches extreme "non-narrativity" in the modern period. In fact, across all modern-day genres and subgenres studied by Biber (1988), there are only three with higher "non-narrative" scores on this dimension than the 1975 score.

Some of the diachronic variation on Dimension 2 can be illustrated by comparing passages from 1725 and 1925. In these samples, bold print marks "narrative" features:

> 1725:
> 6. Food or Sustentation. For the first Year, as **has** been already **observ'd, they** all suck the Dam. After **they** are weaned, the right whales, as is generally **supposed**, live upon some ouzy Matter, which **they** suck up from the Bottom of the Sea. The Triers, that open **them** when dead, **acquaint** me, that **they** never **observed** any Grass, Fish, or any other Sort of Food in the right or Whalebone Whale, but only a grayish soft Clay, which the People call Bole Armoniac; and yet an experienced Whale-man **tells** me, that **he has** seen this Whale in still Weather, skimming on the Surface of the Water, to take in some Sort of reddish Spawn or Brett, as some call it, that at some Times will lie upon the Top of the Water, for a Mile together. (Dudley, 1725, pp. 261–262)

> 1925:
> We **have** not specially illustrated the details in the root, as in general the conditions are similar to those of the stem. A difference, however, that is quite in keeping with general root structure may be **mentioned**. Conifer roots, with exceptions in the case of those produced early by young seedlings, have a structure noticeably more lax and open than that of the stem. The tracheids have larger cavities and comparatively thinner walls, and the zones of summer wood in the growth rings are narrow and abrupt. The resin canals reflect this laxity in the greater breadth of **their** lumina, as compared with those of the stem. In addition the lumen of a horizontal canal in such a position is often found open in the cambial region, and connecting freely with the lumen of a vertical canal as well. (Thomson & Sifton, 1925, p. 75)

Neither of these passages shows large numbers of "narrative" features, but there are about three times as many in the 1725 passage. While early scientific research articles frequently contained substantial amounts of narrative, both of these samples are more in the way of "timeless" descriptions—although there is some reported narrative in the first passage. A greater difference resides in the common use of third-person personal pronouns in the earlier passage, usually referring to the discourse topic, *whales*; the more modern passage, on the other hand, uses full noun forms and demonstrative pronouns to specify and describe various smaller structures of plant anatomy within the roots of the conifer.

A third finding at the linguistic level of analysis concerns a dimension labeled "abstract versus non-abstract information," where "abstract information" is roughly synonymous with passivized style. Three types of passive constructions co-occur with

```
Non-abstract Information

     ---
0     |                         ↑ (Face-to-face conversations = -3.2)
      |
1     |
      |
2     |
      |
3     |
      | 1725
4     |
      |  1675; 1775                          (Official documents)
5     |
      |                               (General academic prose)
6     |
      |
7     |
      | 1825; 1875; 1975
8     |
      | 1925                      (Natural science academic prose)
9    .|.
Abstract Information
```

FIGURE 9.3. Dimension 3 scores for *PTRS* research articles by period. © 1996 Cambridge University Press. Reprinted with the permission of Cambridge University Press.

conjuncts (e.g., *therefore, however*) and adverbial subordinators (e.g., *since, while*) on this dimension. Dimension 3 is of particular interest in the study of scientific research writing since such writing has standardly been characterized as highly passivized (see Ding's contribution to this volume). Diachronic change on Dimension 3 is shown in Figure 9.3.

Figure 9.3 reveals a profound shift in the use of passives and associated features in scientific research writing over the last 300 years—texts in the 17th and 18th centuries cluster at a moderately "abstract" level, while texts in the 19th and 20th centuries are highly "abstract," or passivized. This clustering at two different levels of "abstractness" represents a quite different pattern than that seen on Dimension 1, suggesting dramatic developments in the linguistic history of scientific research writing between 1775 and 1825.

In the text samples given below, the text features comprising this dimension are marked by bold print.

> 1675:
>
> All **being** well **evacuated**, I shook the Engine, **so as that** the little Receiver fell off from the Hail-shot, and stood everywhere close to the skin, **expanded** over the cover of the Glass-plate. Then I had no more to do but to suffer the Air to re-enter into the great Receiver, and this Air pressing upon the little one, kept it so closely fasten'd to its cover, that it was impossible for me to sever them. And I **am assured**, that the Air

enters not into the small Receiver, when **'tis thus applied** upon the skin; for I have often put Gages in them which alwaies kept at the same height, although the Air **was permitted** to repass into the great Receiver. You might also let alone the putting under of the Hail-shot to keep up the little Recipient, because the Air by its Spring would lift it up sufficiently; but then the vacuum would not **be** so perfectly **made.** (Huygens & Papin, 1675, pp. 477–478)

1975:

Lobsters weighing about half a kilogram **were purchased** from a commercial supplier in Maine and **transported** by air express in ice-cooled containers. The animals **were kept** at 11°C in a tank of seawater which **was** continuously **circulated** through a filter of gravel, fibre glass and activated charcoal. The lobsters **were fed** raw fish approximately once per week but **otherwise** the conditions of their confinement complied with those **advised** by the Maine Department of Sea and Shore Fisheries (Goggins 1960).

(b) Anatomy

Observations on the muscles and nerves **were made** either from freshly dissected animals or from limbs **fixed** in Bouin's fixative and then **dissected** in 70% alcohol. The nerves **were studied** by immersing the preparations in solutions of 5% methylene blue **diluted** with seawater (Pantin 1946). Preparations **prepared** in this way **were kept** in a refrigerator at 5°C for several hours. The staining solution **was** then **replaced** with fresh seawater and the preparation **examined** under a binocular microscope. (Macmillan, 1975, p. 4)

Both of these passages come from accounts of experimental methods, and thus should be roughly comparable in terms of rhetorical purpose and situation. The major difference resides in the fact that virtually *all* verbs in the 1975 sample are passivized, as compared to only about one-third of the verbs in the earlier sample; the later passage also has frequent past-participial WH-clause deletions (e.g., *limbs fixed in Bouin's fixative*). These text samples well exemplify the main locus of variation on this dimension—various types of passive constructions of which the agentless passive is the most common—and the results of this analysis can be said to corroborate suggestions that the passive is a high-frequency feature of modern-day scientific research writing.

SYNTHESIS AND INTERPRETATION OF FINDINGS

In this section, I briefly synthesize, and then even more briefly interpret, the results of the rhetorical and MD analyses. First, the profound shift from comparatively "involved" and verbal discourse to highly "informational" and nominal discourse on Dimension 1 of the MD analysis appears to accord closely with the shift from an "author-centered" to an "object-centered" rhetoric.[16] Although by no means neglecting natural phenomena as the ultimate objects of study, early scientific research writing emphasized human actors and their actions, an orientation that was expressed linguistically in relatively "involved" and verbal terms. Over time, however, the author-centered rhetoric was gradually undercut, and it came to serve more specialized rhetorical functions, in more specific textual locations, in texts. By 1875, probably the most frequent use of author-centered rhetoric across all types of articles was to index an author's uncertainty or unique position concerning some part of the text's content. The gradual but progressive evolution away from the author-centered approach has its counterpart in the steady development of research articles toward an "informational" norm at the linguistic level of analysis.[17]

Second, there is also a clear relationship in the *PTRS* between movement toward an object-centered rhetoric and the development of highly "abstract"/passivized language, as seen on Dimension 3. As mentioned above (and see note 16), natural phenomena have always been the central focus of science. But even into the 19th century, the individual directly taking the active role in investigating nature was also given textual prominence, as were that individual's actions. The result was therefore a mixture of emphases—on the individual as agent and on his/her actions as dynamic operations, but also on nature itself as the ultimate object or focus of science. The latter emphasis explains the moderately "abstract"/passivized nature of even early scientific research writing as found in this study; no matter how personalized or author-centered, scientific texts attempt to tell their users ultimately about the object of study, and that object in its various aspects will thus frequently occupy a grammatically or information-structurally prominent place in sentences and clauses, prototypically the grammatical subject/theme position (cf., Halliday, 1985). As the author-centered rhetoric waned, however, the natural objects of study came further to the fore, and their increased prominence was reflected in their ever more common appearance as grammatical subjects and information-structural themes of clauses. These natural objects were, of course, often portrayed as being acted upon, but who was doing the acting—and even the fact that someone *was* still acting—became much less important.

A third common pattern across my two sets of findings is only partly represented in the selection of data reported here; it concerns the use of narrative in *PTRS* texts. Narrative was never a major discourse type in the writing sampled in this study, although it is more prominently represented in the earlier periods, and in certain kinds of text types. A rhetorical-level finding (see Atkinson, 1996) was that experimental reports, in particular, frequently included narrative accounts of how they were performed. Over time, however, actual accounts of experiments—and, in the 20th century, experimental methods as well—became highly circumscribed. The de-emphasis over time of narrative elements is also indicated at the linguistic level of analysis. Results of MD analysis (Figure 9.2) show that, while the group of linguistic features underlying Dimension 2 has never been heavily represented in the *PTRS*, earlier articles were comparatively less *non*-narrative than later articles. These findings—that scores on this dimension were never "narrative" and that they furthermore show a steady and linear trend in the non-narrative direction—dovetail well with the rhetorical-level findings mentioned just above.

Space limitations preclude a detailed interpretation of these synthesized data, but it is only through interpretation that they really start to make sense. Over a decade ago, historians such as Shapin (1984) and Dear (1985) pointed out that the early Royal Society and its leading figure, Robert Boyle, had a rhetorical agenda: to draw adherents to their cause by placing their new approach to natural philosophy in direct opposition to late scholasticism, the then-dominant school of philosophical thought. They did so in part by developing a *rhetoric of immediate experience*. In this system an account of a concrete scientific event, such as an experiment or observation, was elevated to the position of prime authority, thereby offering an alternative to the epistemological touchstone of the scholastics: the words of the ancients (and especially Aristotle). The persuasive power of the new philosophy therefore depended on the success with which its practitioners recreated, in writing or speech, "a definite occurrence happening to a particular person" (Dear, 1985, p. 154).

What these historical accounts don't emphasize, however, and what has since become a dominant focus of historiography on early modern science, is the central rhetorical place played by the *culture of the gentleman*—in the present case the British gentleman (e.g., Daston, 1995; Shapin, 1994; cf., Berman, 1975). According to this view, the British gentleman represented a moral and social ideal—the center around which British society revolved, and from which power flowed. Early modern scientists traded on this conventional image of the gentleman for rhetorical purposes, using it to gain for themselves its special claims to authority, disinterestedness, and moral rectitude. This borrowing was all the more natural since so many scientists were in fact gentlemen. When authors placed themselves in the midst of their texts in the early *PTRS*, they therefore brought to bear a whole social system which powerfully warranted the truth of what they were reporting.

Additional markers of genteel discourse are evident in the 17th- and 18th-century *PTRS*, as well. The very genre form—the polite letter—in which articles most commonly appeared was an important emblem of genteel culture (Klein, 1986; McIntosh, 1986). Likewise, the "dialogic" norms of interaction within the early modern scientific discourse community were modeled on norms of civic conduct that held specifically among gentlemen at this time (Shapin, 1994).

By the beginning of the 19th century, however, an alternative scientific discourse was actively under development. In this approach, the rhetorical focus was on the contextualization of research within communities of researchers and on detailed accounts of scientific technologies, whether abstract as in elaborated sets of methods and procedures or concrete as in minute descriptions of sophisticated instruments. Linguistically, scores on the "abstract"/passivized dimension show a sudden jump, and there are continued steady increases in "informational production"—as we saw before—and in "non-narrativity." All these developments mark the growth of an "object-centered" textual norm—one in which natural phenomena and scientific objects, rather than scientific people, assume increasing importance.

This 19th-century discursive shift ramifies with a period of profound change in the sociocognitive organization in science, whereby a substantially undifferentiated science began to fragment into a constellation of semiautonomous disciplines (e.g., Cannon, 1978; Lepenies, 1988; Miller, 1983, 1986; Rudwick, 1963). In this reorganization, the nascent disciplines developed their identities largely on the basis of distinct investigative methodologies and instrumentation, and (perhaps partly as a result) research began to be viewed in terms of discrete and depersonalized scientific problems. Although this reconceptualization of science occupied (at least in Britain—e.g., Berman, 1975; Morrell, 1971) most of the 19th century, it had begun early on, possibly in connection with an enhanced general interest in quantification and precise measurement (Heilbron, 1993; Porter, 1995; Swijtink, 1987).

The object-centered turn in scientific discourse, first evident in and increasingly represented throughout the 19th century, also marks a textual norm in the 20th. Texts, in fact, grow ever more "informational" and "non-narrative" linguistically, and more impersonal and effaced rhetorically. At the same time, elaborate methods and instrument descriptions become increasingly circumscribed, their rhetorical place appearing to be taken by theoretical descriptions and discussions.

CONCLUSION

In this chapter I have tried to show how differing methodologies of discourse analysis—with their varying theoretical perspectives and analytical techniques—can be used in combination to gain a better understanding of the history of the scientific research article. Obviously, this represents a first attempt rather than a finished product, but by its very unfinished nature this project will hopefully encourage others to follow a similar path. I would claim that it is only by bringing multiple, theoretically independent perspectives to bear—and their findings into direct contact and interaction—that we can get a well-rounded understanding of the history of scientific discourse, as well as other discourses, both past and present.

NOTES

[1] This chapter is based on a presentation given at the 11th World Congress of Applied Linguistics, Jyväskylä, Finland, August 6, 1996. Parts of this chapter have appeared in different form in "The *Philosophical Transactions of the Royal Society of London*, 1675–1975: A sociohistorical discourse analysis" (1996), *Language in Society, 25*, 333–371. I am indebted to Douglas Biber for his continued support and encouragement regarding this research. I would also like to thank Carol Berkenkotter for pointing out insufficiencies in an earlier description of rhetorical analysis (although she bears no responsibility for the present version). The research reported here was supported by NSF grant #BSN-9010893, "Diachronic relations among speech-based and written registers of English" (Douglas Biber and Edward Finegan, principal investigators).

[2] Labov is not alone in calling for the combination of diverse methodologies in the study of social life. Similar proposals have been made by, *inter alia*, DeWalt and Pelto (1985), Eisner and Peshkin (1990), Lazaraton (1995), and Nunan (1992).

[3] The following is a short description of the text corpus analyzed in this research (a fuller

account is given in Atkinson, in press): At the rhetorical level of analysis I examined all original articles occurring in the first bound volumes of the *PTRS* in the years 1675, 1725, 1775, 1825, 1875, 1925, and 1975. For the years 1925 and 1975, I supplemented *PTRS* articles with those in the first bound volumes of the *Proceedings of the Royal Society of London* for those years, since the *Proceedings* had begun by that time to share the function of reporting original research. All told, 211 articles were analyzed at the rhetorical level of analysis. At the linguistic level of analysis, I analyzed 10 articles from each of the seven sampling intervals given above. These articles were selected from the full set of articles subjected to rhetorical analysis using stratified random sampling techniques. This amounted to 70 texts making up a total of approximately 250,000 words.

[4] There is unfortunately no space here to review the pioneering work of researchers on the history of professional discourse in general, and the history of scientific research writing in particular. However, Bazerman's (1988, ch. 3) groundbreaking work on the *PTRS* at least requires mentioning, as both the first extended discourse analysis of this literate scientific landmark, and the main inspiration for the present attempt.

[5] Grammatical/lexical tagging was accomplished using a modified version of Biber's (1988; 1995) program, written in the programming language PL/1. This program relies initially on a 50,624-word online dictionary compiled from the "Brown Corpus" (Francis & Kucera, 1979; 1981), a previously hand-tagged corpus of standard American English. It thus matches forms with entries from that dictionary, using additional context-sensitive disambiguating algorithms (based on Quirk, Greenbaum, Leech, & Svartvik, 1985) where ambiguous usages exist. A second component of the tagger targets lexical items which are not included in the dictionary, using relational information (e.g., "directly preceded by an article and followed directly by a finite verb") to provide probabilistic tags. A second program is used to check items which the tagger does not always tag accurately (e.g., various uses of the lexeme "that"). In my own use of MD analysis I aim for 90 percent tagging accuracy, which is validated by random 100-word checks of tagged texts. The development and implementation of the original tagging program (not including the probabilistic tagger) is described in more detail in Biber (1988, Appendix II).

[6] These findings are reported and interpreted in greater depth elsewhere (Atkinson, 1996; in press). Citations which do not appear in the Reference list refer to articles in the *PTRS*.

[7] For example, 18 separate apparent genre names appear in the titles of *PTRS* articles for 1775. These include: Account, Essay, Experiments, Experiments and Observations, Journal, Letter, Observations, Proposal, and Remarks.

[8] It should be noted that many of the features described here also occurred (but less frequently) in articles from 1675 *not* appearing in letter form.

[9] By 1825, Newton had been president of the Royal Society for 21 years. Unlike all other periods in the 17th and 18th centuries, the Royal Society was at this time substantially focused on a single research agenda—the one set out in the various editions of Newton's major works (see Atkinson, 1993, Appendix 1; Heilbron, 1983).

[10] Bazerman (1988, chap. 3) has also identified this orientation to discourse community in the *PTRS* around this time. Its origins and social significance, however, are by no means clear.

[11] Other indexes of this trend are the large average number of pages devoted to each article (35.5 as opposed to 24.4 in 1825 and 31.3 in 1925), the first appearance of tables of contents for individual articles, and the extremely large numbers of results which quantitative articles report. For the first time, the majority of articles in this period also have explicit top-level discourse structures (Meyer, 1985), which must result at least partly from having to organize the enormous amounts of information contained in articles in some comprehensible way.

[12] Reports in physics and mathematics that are *purely* theoretical pay comparatively little attention to contextualizing the work reported within a larger discourse community. Instead, they frequently start off by simply describing their topic, or by contextualizing it in a cursory way.

[13] The progressive de-emphasis of methods in 20th-century research writing has also been noted by Berkenkotter and Huckin (1995, ch. 2), who attribute it to the desire to foreground new findings.

[14] Some technical information regarding MD analysis will allow for a better understanding of the results discussed here (for more detailed accounts see Atkinson, in press; Biber, 1988): a) Each of the sets of co-occurring features (i.e., factors—see section entitled "The Methods" above) determined via MD analysis has been assigned a functional interpretation based on the linguistic functions shared by the individual features—these interpretations are in the form of dichotomies, e.g., "involved versus informational production"; b) Because different texts—or, more accurately, different text groups (groups of texts in the present study differ in terms of the time period in which they were written, since diachronic change is what is being studied)—can be quantitatively related to one another according to how much or little of the co-occurring sets of features are represented within them, these feature sets and their interpretive labels are construed as continuous "dimensions" on which different kinds of discourse vary, with each dichotomous element of the label representing one "pole" or end of its dimension; c) the actual scores of different text groups on these dimensions are calculated by adding up (with differences in text lengths and absolute frequencies of feature occurrences accounted for statistically) the times a member of a particular feature set occurs in a particular text group. That is, the closer a group of texts is to one pole of the dimensions/scales represented in the figures, the more occurrences there are of individual features making up the particular set of features represented by that dimension. The actual linguistic features represented by each dimension are as follows:

Dimension 1: *Involved versus Informational Production*—private verbs (e.g., *think, feel*), *that*-deletion, contractions, present-tense verbs, second-person personal pronouns, demonstrative pronouns, general emphatics, first-person personal pronouns, pronoun *it*, BE as main verb, causative subordination, discourse particles, indefinite pronouns, general hedges, amplifiers, WH-questions, possibility modals, non-phrasal coordination, WH-clauses, final prepositions // nouns, longer words, prepositions, type/token ratios, attributive adjectives (note: As marked just above via double slash marks, there are actually *two* (oppositional) sets of features represented on this dimension only—scores for individual texts are therefore calculated by summing occurrences of the two sets and then subtracting the second from the first).

Dimension 2: *Narrative versus Non-narrative Concerns*—simple past tense, perfect aspect, public verbs (e.g., *say, report*), third-person personal pronouns.

Dimension 3: *Non-abstract versus Abstract Information*—conjuncts, agentless passives, past-participial WH-clause deletions, other adverbial subordinators, *by* passives.

[15] Thus, the average length of noun phrases in the 1675 passage is 1.7 words, in the 1975 passage 6.9.

[16] The following discussion of this and the next point assumes the existence and widespread cross-historical acceptance of a "cultural model" (Holland & Quinn, 1987) of scientific activi-

ty on the order of: *human → acting on → nature*. Certainly, I would claim that large numbers of scientists implicitly or explicitly adopt this model in accounts of their daily practice (if "acting on" can be taken to include activities such as observing and measuring, as well as dissecting and experimentally manipulating).

[17] The two types of data do not fit together perfectly here—at the rhetorical level, for instance, little substantive change away from an author-centered rhetoric is detectable up to 1775, while at the linguistic level change is unswervingly progressive. But for two very different types of analysis to be sensitive in exactly the same ways to the complex historical developments taking place in scientific writing would not be realistic. As Labov (1971) suggested in a quotation cited early in this chapter, the emphasis must be on the *partial convergence* of data when using multiple methodologies.

REFERENCES

Atkinson, D. (1991). Discourse analysis and written discourse conventions. *Annual Review of Applied Linguistics, 1,* 57–76.

Atkinson, D. (1992). The evolution of medical research writing from 1735 to 1985: The case of the *Edinburgh Medical Journal. Applied Linguistics, 13,* 337–374.

Atkinson, D. (1993). *A historical discourse analysis of scientific research writing from 1675 to 1975: The case of the Philosophical Transactions of the Royal Society of London.* Unpublished doctoral dissertation, University of Southern California.

Atkinson, D. (1996). *The Philosophical Transactions of the Royal Society of London,* 1675–1975: A sociohistorical discourse analysis. *Language in Society, 25,* 333–371.

Atkinson, D. (in press). *Scientific discourse in sociohistorical context: The Philosophical Transactions of the Royal Society of London, 1675–1975.* Hillsdale, NJ: Erlbaum.

Bazerman, C. (1988). *Shaping written knowledge.* Madison, WI: University of Wisconsin Press.

Bazerman, C. (1994). *Constructing experience.* Carbondale, IL: Southern Illinois University Press.

Berkenkotter, C., & Huckin, T.N. (1995). *Genre knowledge in disciplinary communication.* Hillsdale, NJ: Erlbaum.

Berman, M. (1975). "Hegemony" and the amateur tradition in British science. *Journal of Social History, 8,* 30–50.

Biber, D. (1988). *Variation across speech and writing.* New York: Cambridge University Press.

Biber, D. (1995). *Dimensions of register variation: A cross-linguistic comparison.* Cambridge: Cambridge University Press.

Biber, D., & Finegan, E. (Eds.). (1994). *Sociolinguistic perspectives on register.* New York: Oxford University Press.

Biber, D., Finegan, E., & Atkinson, D. (1994). ARCHER and its challenges: Compiling and exploring a representative corpus of historical English registers. In U. Fries, G. Tottie, & P. Schneider (Eds.), *Creating and using English Language corpora* (pp. 1–13). Atlanta, GA: Rudopi.

Brown, R. L., & Herndl, C. G. (1986). An ethnographic study of corporate writing. In B. Couture (Ed.), *Functional approaches to writing: Research perspectives* (pp. 11–28). Norwood, NJ: Ablex.

Cannon, S. F. (1978). *Science in culture: The early Victorian period.* New York: Science History Publications.

Daston, L. (1995). The moral economy of science. *Osiris, 10,* 3–24.

Dear, P. (1985). Totius in verba: Rhetoric and authority in the early Royal Society. *Isis, 76,* 144–161.

DeWalt, B. D., & Pelto, P. J. (1985). Methodology in macro-micro studies. In B. D. DeWalt & P. J. Pelto (Eds.), *Micro and macro levels of analysis in anthropology: Issues in theory and research* (pp. 187–201). Boulder, CO: Westview.

Eisner, E. W., & Peshkin, A. (1990). Introduction. In E. W. Eisner & A. Peshkin (Eds.), *Qualitative inquiry in education: The continuing debate* (pp. 1–14). New York: Teachers College Press.

Ervin-Tripp, S. (1972). On sociolinguistic rules: Alternation and co-occurrence. In J. J. Gumperz & D. Hymes (Eds.), *Directions in sociolinguistics* (pp. 213–250). New York: Holt, Rinehart & Winston.

Ferguson, C. (1994). Dialect, register, and genre: Working assumptions about conventionalization. In D. Biber and E. Finegan (Eds.), *Sociolinguistic perspectives on register* (pp. 15–30). New York: Oxford University Press.

Francis, W. N., & Kucera, H. (1979). *Manual of information to accompany A Standard Corpus of Present-Day American English, for use with digital computers.* Providence, RI: Department of Linguistics, Brown University.

Francis, W. N., & Kucera, H. (1981). *Frequency analysis of English usage: Lexicon and grammar.* Boston: Houghton Mifflin.

Freed, R. C., & Broadhead, G. J. (1987). Discourse communities, sacred texts and institutional norms. *College Composition and Communication, 38,* 154–165.

Goffman, E. (1972). *Frame analysis: An essay on the organization of experience.* Boston: Northeastern University Press.

Halliday, M. A. K. (1985). *An introduction to functional grammar.* London: Edward Arnold.

Halliday, M. A. K. (1988). On the language of physical science. In M. Ghadessy (Ed.), *Registers of written English* (pp. 162–178). London: Pinter.

Heath, S. B. (1983). *Ways with words: Language, life and work in communities and classrooms.* Cambridge: Cambridge University Press.

Heilbron, J. L. (1983). *Physics at the Royal Society during Newton's presidency.* Los Angeles: William Andrews Clark Memorial Library.

Heilbron, J. L. (1993). A mathematicians' mutiny, with morals. In P. Horwich (Ed.), *World changes: Thomas Kuhn and the nature of science* (pp. 81–129). Cambridge, MA: MIT.

Holland, D., & Quinn, N. (1987). *Cultural models in language and thought.* Cambridge: Cambridge University Press.

Huxley, T. H. (1968). On the advisableness of improving scientific knowledge. In *Methods and results: Essays* (pp. 18–41). New York: Greenwood Press. (Original work published 1866)

Klein, L. D. (1986). Berkeley, Shaftesbury, and the meaning of politeness. In O. M. Brack, Jr. (Ed.), *Studies in eighteenth-century culture* (pp. 57–68). Madison, WI: University of Wisconsin Press.

Knorr-Cetina, K. D. (1981). *The manufacture of knowledge: An essay on the constructivist and contextual nature of science.* Oxford: Pergamon.

Kronick, D. A. (1988). Anonymity and identity: Editorial policy in the early scientific journal. *Library Quarterly, 58,* 221–237.

Labov, W. (1971). Methodology. In W. O. Dingwall (Ed.), *A survey of linguistic science* (pp. 413–491). College Park: Linguistics Program, University of Maryland.
Latour, B. (1987). *Science in action*. Cambridge, MA: Harvard University Press.
Latour, B., & Woolgar, S. (1986). *Laboratory life: The construction of scientific facts*. Princeton, NJ: Princeton University Press.
Lauer, J. M., & Asher, J. W. (1988). *Composition research: Empirical designs*. New York: Oxford University Press.
Lazaraton, A. (1995). Qualitative research in applied linguistics: A progress report. *TESOL Quarterly, 29,* 455–472.
Lepenies, W. (1988). *Between science and literature: The rise of sociology.* Cambridge: Cambridge University Press.
MacDonald, S. P. (1994). *Professional academic writing in the humanities and social sciences*. Carbondale, IL: Southern Illinois University Press.
McIntosh, C. (1986). *Common and courtly language: The stylistics of social class in 18th-century English literature*. Philadelphia: University of Pennsylvania Press.
Meyer, B. J. F. (1985). Prose analysis: Purposes, procedures, and problems. In B. K. Britton & J. B. Black (Eds.), *Understanding expository text: A theoretical and practical handbook for analyzing explanatory text* (pp. 11–64). Hillsdale, NJ: Erlbaum.
Miller, C. R. (1984). Genre as social action. *Quarterly Journal of Speech, 70,* 151–167.
Miller, D. P. (1983). Between hostile camps: Sir Humphry Davy's presidency of the Royal Society of London. *British Journal of the History of Science, 16,* 1–47.
Miller, D. P. (1986). The revival of the physical sciences in Britain, 1815–1840. *Osiris, 2,* 107–134.
Morrell, J. B. (1971). Individualism and the structure of British science in 1830. *Historical Studies in the Physical Sciences, 3,* 183–204.
Myers, G. (1990). *Writing biology: Texts in the social construction of scientific knowledge*. Madison, WI: University of Wisconsin Press.
Nunan, D. (1992). *Research methods in language learning*. Cambridge: Cambridge University Press.
Porter, T. M. (1995). *Trust in numbers: The pursuit of objectivity in science and public life*. Princeton, NJ: Princeton University Press.
Quirk, R. J., Greenbaum, S., Leech, G., & Svartvik, J. (1985). *A comprehensive grammar of the English language.* London: Longman.
Rogoff, B. (1990). *Apprenticeship in thinking: Cognitive development in social context*. New York: Oxford University Press.
Rudwick, M. J. S. (1963). The foundation of the Geological Society of London. *British Journal of the History of Science, 1,* 325–355.
Schriffrin, D. (1994). *Approaches to discourse*. Oxford: Blackwell.
Shapin, S. (1984). Pump and circumstance: Robert Boyle's literary technology. *Social Studies of Science, 14,* 481–520.
Shapin, S. (1994). *A social history of truth: Civility and science in seventeenth-century England*. Chicago: University of Chicago Press.
Spradley, J. P. (1980). *Participant observation*. New York: Holt, Rinehart and Winston.
Stubbs, M. (1996). *Text and corpus analysis*. Oxford: Blackwell.
Swales, J. (1990). *Genre analysis: English in academic and research settings*. Cambridge: Cambridge University Press.
Swijtink, Z. J. (1987). The objectification of observation. In L. Krüger, L. J. Daston, & M.

Heidelberger (Eds.), *The probabilistic revolution* (Vol. 1, pp. 261–285). Cambridge, MA: MIT.

Tannen, D. (1990). Discourse analysis: The excitement of diversity. *Text, 10,* 109–111.

Tarone, E., Dwyer, S., Gillette, S., & Icke, V. (1981). On the use of the passive in two astrophysics journal papers. *ESP Journal, 1,* 123–140.

Van Dijk, T. A. (1985). *Handbook of discourse analysis.* San Diego: Academic.

Vande Kopple, W. J. (1994). Some characteristics and functions of grammatical subjects in scientific discourse. *Written Communication, 11,* 534–564.

Wells, R. (1960). Nominal and verbal style. In T. Sebeok (Ed.), *Style in language* (pp. 213–220). Cambridge, MA: MIT.

10

Metaphor in the Rhetoric of Scientific Discourse

Richard D. Johnson-Sheehan
University of New Mexico

Since the Enlightenment, a cherished notion in Western culture has been that scientific discourse is ideally transparent and plain, devoid of the rhetorical and metaphorical maneuvers that are common in more humanistic forms of discourse like literature and poetry. Scientific discourse, it is often assumed, ideally provides a clear window into the workings of nature, helping us observe an "outside" reality that is often clouded by our own all-too-human subjectivities. Moreover, there has been a recurring backlash against figurative language in scientific discourse, originating with Enlightenment scholars like Hobbes, Locke, and Sprat. In the *Leviathan*, Hobbes (1651/1958) specifically accuses metaphor of being a "deception" in philosophy and science, because metaphors use words "in other sense than that they are ordained for, and thereby deceive others" (p. 166). Locke (1690/1961), in his *Essay Concerning Human Understanding*, argues that rhetoric and metaphor are used to "insinuate wrong ideas, move the passions, and thereby mislead the judgement, and so are perfect cheat" (p. 105). Locke then attempts to banish rhetoric from scientific discourse by arguing that it is an instrument of "error and deceit" (p. 106). Finally, in his *History of the Royal Society*, Sprat (1667/1959) applauds his fellow members of the Royal Society for reforming language to avoid "this vicious abundance of Phrase, this trick of Metaphor, this volubility of Tongue, which makes so great a noise in the World" (p. 112). For Hobbes, Locke, Sprat and many of their descendants, scientific discourse (and rationality for that matter) is defined by the *absence* of figurative language. Science, so the myth goes, only deals in propositions and concepts based on words that are exactly defined; consequently, figurative language is undesirable because it is assumed to be a warping or twisting of language, better left to poets and politicians. Even today, the repeated calls for a universal scientific language that only uses "simple" or "concrete" terms reflects the prejudice of a scientific community that still believes scientific discourse can be non-rhetorical and non-figurative.

The last few decades, however, have seen a revival of interest in scientific metaphor not only among rhetorical theorists but also among scientists (Bohm & Peat, 1987; Jones, 1982; March, 1970). My intention in this chapter is to discuss this re-emergence of interest in scientific metaphor and figurative language, demonstrating not only the importance of metaphor to scientific discourse but also its centrality in the invention of scientific theories. Reviewing some of the prominent work that has been done in this area, I hope to outline some of the potential paths that research in scientific metaphor might follow while loosely staking out the possible limits of this research in the growing field of rhetoric of science. Toward the end of the chapter, I will discuss metaphorical analysis as a means for studying and illuminating the rhetoric of scientific texts.

THE IMPORTANCE OF METAPHOR IN SCIENCE

To say that scientific discourse is inherently metaphorical might suggest a falsehood to the staunch realist. Perhaps that person might even take such a suggestion to be a degradation of science itself. And yet, as Richards (1936) said of metaphor in all language, "We cannot get through three sentences of ordinary fluid discourse without it. ... Even in the rigid language of the settled sciences we do not eliminate or prevent it without great difficulty" (p. 89). Richards is far too modest. In fact, it is impossible to avoid using metaphors in scientific discourse. Some of the more obvious examples of metaphor in science include statements like William Harvey's "the heart is a pump" or Thomas Young's "light is a wave." Some less obvious examples might include Copernicus' "the universe is heliocentric" or Darwin's "natural selection," or even Einstein's "time and space are relative." Moreover, new scientific terms, like "gravity," "double helix," or "aether wind" tend to emerge first as metaphors employed to fill a semantic gap when a new concept emerges. Clearly disproving prejudices that scientific discourse is non-figural, the lexicon of science is wholly saturated with living and dead metaphors. And new metaphors are being created constantly as scientists attempt to use language to make sense of the natural phenomena they experience. To quote again from Richards (1936), "The pretense to do without metaphor is never more than a bluff waiting to be called" (p. 89). Ironically, despite the recurring attacks on metaphor since the Enlightenment, this seems doubly true in scientific discourse.

My intention, however, is not to offer yet another sermon in which scientific metaphors are identified for the unbelievers. The pervasiveness of metaphor has been touched upon so often by theorists that it need not be addressed anew here. Instead, I propose to first discuss an issue that has not been resolved to anyone's comfort: the extent of metaphor's influence in scientific theories and beliefs. Richards (1936) was among the first to argue that "*Thought* is metaphoric, and proceeds by comparison, and the metaphors of language derive therefrom" (p. 90). More recently, theorists like Lakoff and Johnson (1981) have argued, "Our ordinary conceptual system, in terms of which we both think and act, is fundamentally metaphorical in nature" (p. 287). They point out that basic cognitive concepts like "up" and "down" are rooted in metaphors. More important to scientists, though, Lakoff and Johnson demonstrate that concepts like "time" and "space," which play a central role in any post-Newtonian theory of physics, derive their meaning from metaphors. For example, Lakoff and Johnson argue that metaphors like "time is money" impel us to conceptualize temporal issues in quantitative terms (pp. 290–291). Of course, one of Newton's great achievements was to recognize the importance of an absolute, quantifiable concept of time (and space) to algebraic or calculus-based theories of mechanics and astronomy. Before Newton, time and space were conceptualized in far more qualitative and less absolute terms. In all, like "time," many of the fundamental concepts on which physics, chemistry, and other sciences are based—like "force," "energy," "cell," "space," or "molecule"— are only definable in terms of metaphor. Indeed, they were more or less dead metaphors in the first place.

But if Richards, and Lakoff and Johnson are right, it seems we are compelled to acknowledge that even the scientific thought and the conceptual systems on which scientific beliefs are founded are essentially metaphoric in nature. As most anyone can guess, this reversal of fortune for scientific metaphor is hard to accept for those people calling for "concrete" non-figural scientific language. The question that then faces us is whether scientific beliefs, not just the lexicon of science, is essentially metaphoric or at least structured conceptually by metaphors. To this question, scholars like Arbib and Hesse (1986) answer an emphatic "yes." They write, "all language is metaphorical" (p. 150) and that "scientific revolutions are, in fact, metaphoric revolutions, and theoretical explanation should be seen as metaphoric redescription of the domain of phenomena" (p. 156). In making these sweeping statements, Arbib and Hesse claim that they are arguing for a "nonliteralist" interpretation of not only language but also reality (p. 150). They suggest that as we come to terms with a reality that is ultimately indeterminate, our terminology and beliefs must also be able to shift gradually at the semantic level. "General terms are necessary," Arbib and Hesse concede, "but their use entails some loss of information about the world. ... in practice we allow that language works by capturing approximate meanings, that degrees of similarity and difference are sufficiently accessible to perception to avoid confusion in ordinary language" (p. 152). Consequently, Arbib and Hesse claim, the malleable nature of metaphor suits attempts to characterize an indeterminate reality because metaphor is "intersubjectively understood" and thus open to such subtle shifts of meaning in the scientific lexicon (p. 154).

Arbib and Hesse, and others who adopt this strong position toward the role of metaphor in science, leave themselves open to some rather fundamental epistemological challenges. After all, if "all language is metaphorical," as they claim, then metaphor itself seems to lose its own definition. In other words, once metaphor is universalized to mean "all language," then it is hard to speak of any unique, identifiable status for metaphor in a study of scientific discourse. For this reason, most scholars of scientific metaphor step back from this abyss. Peterfreund (1994), for example, argues that the scientific lexicon is primarily made up of metaphor, metonymy, and other tropes (p. 73). Peterfreund suggests that whereas metaphors lead to dynamic transferrals of meaning that might spark revolutions in science, metonymy is far more static because it uses literal words to stand for particular concepts (p. 64). Peterfreund points out that whereas the nature of metaphor is to seek out change, the nature of metonymy is to maintain stability and the status quo. "It may be the case," he argues, "that the movement from revolution to normal science is a movement from metaphor to metonymy. … In metaphor, or pure theory, the relationship of the two terms of the equation is transferential and fluid; in metonymy, the relationship is substantive and specified as that of effect for cause or cause for effect" (p. 73). Indeed, most theorists of metaphor adopt a weak view of metaphor similar to Peterfreund's. They typically maintain some kind of division between metaphor (figurative language) and literal language. Metaphor is then seen to be an agent of change, leading to shifts in theories and the literal language that scientists use to describe reality.

Nevertheless, that metaphor is fundamentally related to change in scientific language and beliefs seems to be a consistent conclusion among rhetorical scholars. On one hand, for scholars like Arbib and Hesse, scientific discourse is wholly dominated by metaphors—they actively structure the thought patterns of scientists while offering a lexicon that shifts gradually as the concepts shift. Scholars like Peterfreund, on the other hand, try to assign metaphor a limited role as an agent of change. Metaphors may bring about change, but in the end they only modify the literal lexicon on which scientific discourse is based. For scholars like Peterfreund, other forms of language, like metonymy or even so-called "literal language," offer a more static connection with the observable world.

WHAT METAPHORS DO

Assuming that the pervasiveness of metaphor is an unresolved issue, let us look closer at the role metaphors play in scientific discourse. Metaphor, as Burke (1954) points out, invites hearers to change perspective or adopt a particular point of view, effectively guiding how they "see" and make sense of their experiences (p. 504). If Burke is correct, then what we take to be "reality," including scientists' versions of reality, could be viewed simply as a world comprehended through a selective perspective. Indeed, most theorists agree that metaphors obligate us to interpret our experiences with reality *in terms of* something else. Black (1977) highlights the importance of metaphor in guiding how one perceives reality when he writes, "But it is no longer surprising if one believes that the world is *necessarily a world under a certain description*—or a world seen from a certain perspective. Some metaphors can create such a perspective" (p. 454).

So, what do metaphors do? To put it concisely, metaphor is a device for guiding or changing perspective. When speakers use metaphor, they invite their hearers to see something from a different point of view. Burke (1954) proposes that metaphor urges hearers to consider one thing (A) from the point of view of something else (B), using "B as a *perspective* on A" (p. 504). When a speaker changes perspective by employing a metaphor, Burke argues, he can then discourse about something "in terms of" something else (p. 504). For example, suppose someone were to say "nature is a garden," a metaphor common in pre-Enlightenment natural philosophy. A hearer, identifying a potential metaphor, might interpret the speaker to mean that nature is something humans need to control and tend. Perhaps the hearer might reconceptualize the Earth as the dominable source for food in which humans are of central importance. Maybe the hearer might even come to believe that humans are gardeners or farmers appointed by the gods to use nature as they see fit. Now, suppose someone were to say, "nature is a machine" as Kepler, Descartes, and Newton did. This change in metaphor invites the hearer to change perspective. She might now view the motions of the Earth in terms of the inner workings of grand mechanism. From this perspective, the heart becomes a pump, the brain becomes a computer, the Earth spins on its axis, things move according to rigid laws and forces, and God becomes a "divine engineer" or "clockmaker." My point is that metaphors guide a hearer's perspective—a change in metaphor is a change in perspective.

Similarly, in scientific discourse, metaphors can also be used to guide and change perspective. For example, consider Newton's (1730/1952) use of the metaphor "Are not the Rays of Light very small Bodies emitted from shining Substances" on which he developed his corpuscular theories of light (p. 370). Interpreted from Newton's perspective, the phenomenon of light is to be described *in terms* that scientists typically associate with particles. Using this metaphor, Newton explained the refraction of light by arguing that larger objects act on particles of light "at a distance," bending the stream of particles much as gravity bends the path of larger objects (p. 371). According to Newton, the gravitation involved in refraction explains why rays of light particles are bent by a glass surface or around the corner of a wall. Newton explained the colors of a prism by suggesting that the glass of the prism sorts white light into its constituent colors by refracting light particles of different color in different degrees (p. 20). He theorized that the polarization of light occurs because light particles have "two sides" much as a magnet has two poles (p. 360). Polarized light, Newton argued, is created when the "sides" (i.e. the poles) of the particles in rays of light are in alignment with one another. Even the wave-like properties of light that ultimately undermined Newton's corpuscular theories are explained in particle terms. Newton writes "as Stones by falling upon Water put the Water into an undulating motion, and all Bodies by percussion excite vibrations in the Air; so the Rays of Light, by impinging on any refracting or reflecting Surface, excite vibrations in the refracting or reflecting Medium or Substance" (p. 280). Indeed, the metaphor "light is a corpuscle" provided a useful perspective that Newton employed to conceptualize and discourse about light.

A century later, however, a switch in perspective, or metaphor, by Young radically changed theories of light. Young employed a very different metaphor, "light is the undulation of an elastic medium," to invent his "wave" theory of light (as cited in Wood, 1954, p. 160). Seeing the phenomenon of light from a quite different perspective offered by this metaphor, Young argued that light is a wave traveling through ether, a nearly undetectable substance that filled all empty space. Much as waves travel through mediums of water or air, he suggested, light is a wave in ether. Describing light *in terms of* a wave, Young argued that light has a "wavelength," "frequency," and "amplitude" much like waves in air, water, or a violin string. Essentially, the metaphor, "light is a wave," invited Young and his followers to describe light through a wave-based theory that used terms associated with waves, not particles.

Indeed, in a 1801 lecture by Young to the Royal Society, we see a comprehensive theory developing, based on this new perspective. After setting out his main hypothesis that light is a wave, Young offers four further hypotheses that are based on the "light is a wave" metaphor:

1. A luminiferous ether pervades the universe, rare and elastic in a high degree.
2. Undulations are excited in this ether whenever a body becomes luminous.
3. The Sensation of different Colours depends on the different frequency of Vibrations excited by Light in the Retina.
4. All material Bodies have an Attraction for the ethereal Medium, by means of which it is accumulated within their substance, and for a small Distance around them, in a state of greater Density, but not of greater Elasticity. (as cited in Wood, p. 160)

In these hypotheses, we see Young inventing a point-by-point contradiction of Newton's corpuscular theories of light by reconceptualizing the phenomenon of light in terms of waves—from a different perspective. Young's use of a different metaphor invites him to invent a very different theory that describes the behavior of light. To sum up at this point, in Newton's and Young's efforts to develop theories of light, we see clearly how competing metaphors provide differing perspectives from which scientists conceptualize phenomena differently. Moreover, we see how a change in root metaphor—in this case a change from "light is a particle" to "light is a wave"—changes the point of view from which one discourses about light.

HOW SCIENTIFIC METAPHORS WORK

Interestingly, one of the great debates concerning metaphor (i.e. how metaphors work) has not largely impacted discussions about how scientific metaphors work. Since the study of scientific metaphor began a few decades ago, the great majority of theorists have subserviently followed the "interaction" or "tension" view of metaphor developed by Richards and elaborated by Black. The interaction view suggests that the two parts of a metaphor, the tenor and vehicle, create a semantic tension in the mind of the hearer. The result of this tension is a new meaning, insight, or cognitive content that transcends the so-called "literal" meanings of the two words. For instance, consider one of my previous examples from Young, "light is a wave." Proponents of the interaction view would suggest that the vehicle "wave" is interacting with the tenor "light" to provoke a new meaning in the mind of the hearer. The vehicle, as Black (1981) argues, then serves as a "filter" through which the "metaphor selects, emphasizes, suppresses, and organizes features of the principle subject by implying statements about it that normally apply to the subsidiary subject" (p. 78). In other words, the word "wave" now serves as a filter through which one restructures the meaning of "light."

Over the last half century, many scholars have applied Richards and Black's interaction theories to scientific metaphors. One of the first comprehensive discussions of scientific metaphors was offered by Hesse (1966) in her book *Models and Analogies in Science*. In this important treatise, Hesse uses Black's description of the "interaction" view of metaphor to explain how metaphors bring about changes in scientific "models." Lightly touching on Kuhn's (1970) notions of paradigm and revolution, Hesse argues that metaphors provide the structural foundations for scientific theories. She writes, "Metaphors ... may be seen as correcting and replacing literal descriptions of the same system, so that the literal descriptions are discarded as inadequate and false" (p. 174). Two decades later Hesse, with Arbib, built upon her original arguments by strengthening connections to Kuhn's work and employing the vocabulary of social cognition. Arbib and Hesse (1986) argue that new metaphors compel scientists to "see" phenomena differently, causing so-called "literal" terms to "shift toward the metaphorical meaning" (p. 156). Metaphors, Arbib and Hesse claim, alter scientists' "socially constructed" cognitive schema, because they are "incompatible" with current scientific paradigms (p. 2). They write:

> To use Kuhnian terminology, in the development of science "normal science seeks to reduce instability of meaning and inconsistency and to evolve logically connected theories; revolutionary science makes metaphoric leaps that are creative of new meanings and applications that may constitute genuine theoretical progress. (p. 157)

Therefore, Arbib and Hesse argue, metaphors become the impetus for both stability and change in scientific communities, because metaphors act as a structural basis for new paradigms while also periodically undermining established scientific paradigms.

MacCormac (1976) in *Metaphor and Myth in Science and Religion* offers a broader and less defined understanding of the role of metaphor in science. Relying on Kuhn's ideas of revolution, he writes:

> The very nature of science, however, is such that scientists need the metaphor as a bridge between old and new theories. Only by some such device as a metaphor is it possible to develop new meanings in new theories that are intelligible. ... A metaphor results from the juxtaposition of the old and new. Our familiarity with the old usage allows us to recognize the terms as meaningful while the new usage suggests a different way of comprehending its meaning. (p. 36)

MacCormac argues that the "tension" created by new scientific metaphors "shock" scientists into seeing reality differently (pp. 73–86). In some cases, he claims, metaphors become root metaphors on which scientists develop "myths," much as religious myths are developed and ultimately taken (or mistaken) for literal truth. Indeed, MacCormac suggests that the history of science is characterized by the misguided attempts of scientists to transform their myths into literal truth. He writes, "By accepting a theoretical description as a literal one, men have over and over transformed a speculative and hypothetical understanding into a myth" (p. xiii). Eventually, though, falling back on Kuhn, MacCormac argues that these myths run up against unexplainable anomalies, forcing scientists use new metaphors to help them make bridges into different theories.

One of the more interesting adaptations of the interaction view to scientific discourse is found in Gerhardt and Russell's *Metaphoric Process: The Creation of Scientific and Religious Understanding* (1984). Gerhardt and Russell argue that metaphors "fold a map of a world of meanings," thus creating a metaphor that "reforms fields of meanings themselves" (p. 113). In scientific discourse, Gerhardt and Russell suggest, this map folding results in a "flash" of cognitive insight. They write, "It is here that the words 'Eureka, I have it!' are spoken. At this point an 'ontological flash' occurs" (p. 114). The metaphor, according to Gerhardt and Russell, then causes a "structural change which demands that other meanings and understandings have to be changed in the wake of the metaphor" (p. 119).

Difficulties with Kuhn's theories aside, the heavy reliance by most rhetoricians of science on interactionist views of metaphor has probably artificially stunted research into scientific metaphor. The problem with the interaction view, as discussed separately by Davidson (1979) and Searle (1979), is the belief that the *interaction* of words in a metaphor creates flashes of insight, non-literal cognitive meanings, mental shocks, or ontological flashes. Quite differently, Davidson and Searle point out that metaphors work the same way as any other literal phrase—their meanings are invented through acts of invention and interpretation by the hearer. So, to suggest that a "flash" or a "shock" sets metaphors apart from other forms of language is to deny that metaphors make up the bulk of our everyday discourse. Indeed, if Richards is right when he claims that a person cannot speak three sentences without using a metaphor, then speakers and hearers would be experiencing shocks and flashes almost continuously.

In studies of scientific discourse, the problems with the interaction view of metaphor are greatly magnified. After all, many of the metaphors on which scientific theories are based, like "light is a wave" or "nature is a machine," have been around for millennia in various forms. If metaphors create shocking episodes or "Eureka, I have it" moments, then they have consistently missed their mark for centuries. Or, if metaphors provide insights, we need to ask a sticky epistemological question—insights into what? I do not mean to suggest that discussions about how scientific metaphors "work" are not helpful. Indeed, one potential direction for research on scientific metaphor might be to work through the criticisms of the interaction theory of metaphor by studying metaphors found in scientific texts. After all, unlike the majority of metaphors in literature, poetry, and politics, scientists typically do not intend for the meaning of their metaphors to be ambiguous, subtle, or obscure. Perhaps this desire to be unambiguous on scientists' part would make scientific metaphors ideal candidates for studying how metaphors work. Another possibility would be to try employing theories of metaphor other than Black's interaction theory—like those of Searle, Goodman, Henle, or Ricoeur—to illuminate how metaphors work in scientific texts.

HOW SCIENTIFIC METAPHORS ARE USED

Perhaps the challenge facing theorists of scientific metaphor is not in developing an answer to the question "How do metaphors work?" Indeed, this might be the wrong question to ask in the first place. If, as Davidson (1979) points out, all metaphors make meaning much as other utterances make meaning—through an "interplay of inventive construction and inventive construal"—then focusing on how metaphors "work" might be inviting us to concentrate on the wrong issue (p. 29). Consequently, to escape the epistemological problems associated with the interaction view, a few rhetoricians of science have begun addressing scientific metaphor from a different direction. Instead of asking how scientific metaphors "work," these scholars are now researching how metaphors are "used" by scientists in theory building (Cantor, 1987; Ivie, 1987; Papin, 1992). In doing so, these rhetoricians posit that metaphors, regardless of how they work, are used by scientists to *invent* scientific theories.

Ironically, this so-called "new" view of scientific metaphor merely follows up something Burke (1954) wrote many years ago. He asks, in *Permanence and Change,* "As the documents of science pile up, are we not coming to see that the whole works of scientific research, even entire *schools*, are hardly more than the patient repetition, in all its ramifications, of a fertile metaphor?" (p. 95). Burke seems to be suggesting that scientists return over and over to "fertile" metaphors for guidance in fleshing out new directions in research. And, as Burke implies, when we concentrate on how scientists "use" metaphors, we begin to see that metaphors are often used as a site for the *invention* of scientific theories. In other words, metaphors like "light is a wave" or "nature is a machine" can serve as motivating perspectives from which scientists continually expand and further invent new ways to conceptualize natural phenomena. For example, Johanes Kepler while composing his masterwork *Astronomia Nova* (1609) wrote of his effort, "My aim is to show that the celestial machine is to be likened not to a divine organism but rather to a clockwork" (as cited in Holton, 1988, p. 56). In this important statement, Kepler identifies two fundamental metaphors that were used to invent very different theories of nature. The first, "nature is a machine," eventually became the basis of "mechanistic" theories that dominated the Enlightenment science (and still dominate much of our interpretations of reality today). The second, "nature is an organism" is the basis of the neo-Aristotelian science, a view vehemently rejected by Enlightenment scientists like Hobbes and Descartes. In making this statement, Kepler clearly demonstrated that he intended to reinvent science from a different perspective, using a quite different metaphor.

A change in root metaphor, however, is only the beginning of a scientific movement. Less than a half century after Kepler, Descartes, one of the architects of Enlightenment, recounts how he expanded the metaphor "nature is a machine" into a philosophy of science. He writes in *Principia Philosophiae* (1644):

> And I have been greatly helped by considering machines. The only difference I can see between machines and natural objects is that the workings of machines are mostly carried out by apparatus large enough to be readily perceptible by the senses... whereas natural processes almost always depend on parts so small that they utterly elude our senses. (as cited in Kearney, 1971, p. 156)

These quotes by Kepler and Descartes show what often happens during the invention of scientific beliefs and theories. "Fertile" metaphors, as Burke calls them, are used by scientists and philosophers to cast new light on previous beliefs about known phenomena. Then, as evidenced in the comments of Kepler and Descartes, scientists continue to return over and over to a particular metaphor to invent coherent theories that explain their experiences with reality. In many cases, as Burke argues, scientists who adopt a shared metaphor—like mechanism or organicism—band together into schools or communities that collectively seek to broaden and flesh out the implications of a fertile metaphor. Indeed, the Royal Society of London can be seen as the banding together of scientists interested in promoting mechanism as a new root metaphor for conceptualizing nature.

Curiously, some of the impetus for this new direction in research on scientific metaphor comes from scientists themselves. Physicists Bohm and Peat (1987) in *Science, Order, and Creativity*, argue that metaphors are an inevitable result of scientific research. Indeed, they suggest that "metaphorical play" is necessary to creativity in science, because it leads to the creation of new theories. The mere activity of doing science, Bohm and Peat claim, brings metaphors forward. Then scientists use the implications highlighted by these new metaphors to develop new theories. They write:

> Within this play it is not to be taken for granted that new things must always be different or that they can never in any significant way be related to what came before. Indeed, it might be suggested that the more different things are, the greater may be the importance in seeing how they are similar, and likewise, the more similar things are, the greater may be the value in perceiving their differences. (p. 49)

Bohm and Peat stress that the *pluralism* created by the give and take among different scientific beliefs and theories urges concepts to play with one another metaphorically in ways that gradually change scientists' perspectives and lead to new theories (p. 49). Then scientists use the similarities and differences among scientific concepts to reconceptualize known facts from a different perspective. This use of metaphor in science, Bohm and Peat claim, is the basis of scientific creativity. Other books by scientists, including Jones' *Physics as Metaphor* (1982) and March's *Physics for Poets* (1970), make similar claims about the importance of figurative language in scientific creativity and theory-building.

Viewing metaphor as a source for rhetorical invention, however, changes our view of the role of scientists in theory building. Instead of being "discoverers" of truth, scientists are actually *interpreters* of fertile metaphors, reconceptualizing phenomena from particular perspectives urged by these metaphors. Or, to put it another way, the purpose of science and scientific discourse is not to lay bare the skeleton of reality; rather, scientists use various perspectives offered by metaphors to make sense of the seemingly chaotic reality that is always changing around us. If, as Black suggests, the world is always a world seen through a selective perspective, then scientists can be seen as interpreters of reality, using particular perspectives to invent theories that explain our experiences with natural phenomena.

CONCLUSION

I would like to conclude by pointing out that little is actually known about scientific metaphors and how they work their magic. Perhaps our lack of knowledge is due to the slippery nature of metaphor itself. If we cannot make up our minds about how "normal" metaphors work or how they are used, then how are we to resolve these issues with scientific metaphors? On the other hand, perhaps research on scientific metaphors might actually be more helpful in understanding metaphor in general. As I mentioned earlier, scientific metaphors are usually not intended to be obscure like poetic metaphors, and they are often situated in very straightforward rhetorical and historical contexts. In many ways, these straightforward contexts make them much easier to study than metaphors found in poetry and literature, which often appear within limited contexts and momentary historical backdrops. Scientific metaphors typically live long lives and can be studied in many different historical contexts.

I would argue further that less theorizing needs to be done about scientific metaphors, and more metaphorical analyses should be conducted on actual scientific texts. Typically, articles on scientific metaphor pluck a particular metaphor out of its context in order to demonstrate that the phrase is indeed a metaphor. Although it is certainly interesting to look closely at a particular scientific metaphor, removing it from its context is like an entomologist removing a beetle from its habitat and observing it in an empty tank at his office. He can watch it, prod it, dissect it, but ultimately he will be able to tell us only limited things about the beetle. Mostly, he would be able to confirm that, yes, it is a beetle. Without observing the beetle in its habitat, as any good entomologist will tell you, he and we only know a small part of the story. Likewise, a researcher who concentrates on a particular metaphor without looking closely at the context in which it was used will likely not see how the metaphor urges a larger shift in perspective by a specific scientist and the broader scientific community. My hope is that more metaphorical analyses would help us study how various metaphors work in various contexts, eventually allowing us to draw some informed conclusions that are based on the commonalties we observe among these metaphors.

Finally, if metaphors do organize and structure our thoughts and our interpretations of reality, then studying scientific metaphors should provide us great insight into not only scientific discourse but also science itself. The greatest benefit in studying scientific metaphors might ultimately be in identifying their limitations or recognizing their untapped potential as sites for invention.

REFERENCES

Arbib, M., & Hesse, M. (1986). *The construction of reality*. Cambridge: Cambridge University Press.

Black, M. (1981). Metaphor. In M. Johnson (Ed.), *Philosophical perspectives on metaphor* (pp. 63–82). Minneapolis, MN: University of Minnesota Press.

Black, M. (1977). More about metaphor. *Dialectica, 3,* 431–457.

Bohm, D., & Peat, D. (1987). *Science, order, and creativity*. New York: Bantam.

Burke, K. (1945). *A grammar of motives*. Berkeley, CA: University of California Press.

Burke, K. (1954). *Permanence and change*. New York: Bobbs-Merrill.

Cantor, G. (1987). Weighing light: The role of metaphor in eighteenth-century optical discourse. In A. Benjamin, G. Cantor, & J. Christie (Eds.), *The Figurative and Literal* (pp. 124–146). Manchester, England: Manchester University Press.

Davidson, D. (1979). What metaphors mean. In S. Sacks (Ed.), *On metaphor* (pp. 29–45). Chicago: University of Chicago Press.

Gerhardt, M., & Russell, A. (1984). *Metaphoric process: The creation of scientific and religious understanding*. Fort Worth, TX: Texas Christian University Press.

Hesse, M. (1966). *Models and analogies in science*. Notre Dame, IN: Notre Dame University Press.

Hobbes, T. (1958). *Leviathan*. In F. Woodbridge (Ed.), *Hobbes Selections* (pp. 134–418). New York: Scribners. (Original work published 1651)

Holton, G. (1988). *Thematic origins of scientific thought*. Cambridge, MA: Harvard University Press.

Ivie, R. (1987). Metaphor and the rhetorical invention of cold war "Idealists." *Communication Monographs, 54,* 165–182.

Jones, R. (1982). *Physics as metaphor*. Minneapolis, MN: University of Minnesota Press.

Kearney, H. (1971). *Science and change, 1500-1700*. New York: McGraw-Hill.

Kuhn, T. (1970). *The structure of scientific revolutions* (2nd ed.). Chicago: University of Chicago Press.

Lakoff, G., & Johnson, M. (1981). Conceptual metaphor in everyday usage. In M. Johnson (Ed.), *Philosophical perspectives on metaphor* (pp. 286–325). Minneapolis, MN: University of Minnesota Press.

Locke, J. (1961). *Essay concerning human understanding*. In J. W. Yolton (Ed.), *An essay concerning human understanding*. London: Dent. (Original work published 1690.)

MacCormac, E. (1976). *Metaphor and myth in science and religion*. Durham, NC: Duke University Press.

March, R. (1970). *Physics for poets*. New York: McGraw-Hill.

Newton, I. (1952). *Opticks; or, a treatise of the reflections, refractions, inflections & colours of light*. New York: Dover. (Reprinted from *Opticks*, by I. Newton, 1730, 4th ed., London: W. Innys)

Papin, L. (1992). This is not a universe: Metaphor, language, and representation. *PMLA, 107,* 1253–1265.

Peterfreund, S. (1994). Scientific models in optics: From metaphor to metonym and back. *Journal of the History of Ideas, 55,* 59–73.

Richards, I. A. (1936). *The philosophy of rhetoric*. Oxford: Oxford University Press.

Searle, J. (1979). *Expression and meaning*. Cambridge: Cambridge University Press.

Sprat, T. (1959). *History of the Royal society*. St. Louis, MO: Washington University Press. (Reprinted from *The history of the Royal-society of London*, by T. Sprat, 1667, London: J. Martyn)

Wood, A. (1954). *Thomas Young, natural philosopher 1773–1829*. Cambridge: Cambridge University Press.

11

The Language of Classifying in Introductory Science Texts*

Steven Darian
Rutgers University

> The ultimate paradox of the intellect is that
> of the part trying to understand the whole.

In this chapter, I would like to explore the syntactic, lexical, and graphical features of classifying, as they appear in two university-level introductory science texts: one in biology (Starr, 1991) and one in chemistry (Hein, et al, 1993). We will also look at several other markers and features of classifying, including plurals, relaters, and quantifiers. We will close with a section on problems in classifying for learning and teaching, for both native and nonnative speakers.

We will begin with a few observations on the role, or functions, of classifying. In addition, we will briefly examine the concept of classifying and its development in Western thought, especially as it bears on our analysis of introductory science texts. This section will also contain a short treatment of induction and deduction, which is central to the classifying process. It also explores the graphical representation of classifications, especially classification trees, sentence lists, and pictures. Quotes from the biology text will be designated as **B**, those from the chemistry text as **C**.

THE ROLE, OR FUNCTIONS, OF CLASSIFYING

Far from a dry mechanical operation, classifying is one of the most important mental skills we have for understanding our world (Crowson, 1970); perhaps, as geographer David Harvey says, it is the basic procedure for finding order in that world (1969). It is also critical to our survival. Categorization, in the words of Bruner, Goodnow, and Austin (1956), "serves to cut down the diversity of objects and events that must be dealt with uniquely by an organism of limited capacities" (p. 235).

This need to classify, suggests biologist Alec Panchen (1992), "also applies to other animals and is by no means confined to animals closely related to man. It is literally *vital* for any animal," Panchen continues, "to have a series of metaphorical compartments in which to place perceived phenomena—food, drink, shelter, danger, own species" (p. 1).

In addition, urges psychologist William Estes (1994), "classification is basic to all of our intellectual activities" (p. 4) and is a major concept in contemporary thought (Ellen, 1979). It is central to text-based research, which tends to be a constant interaction between the preliminary outline (classifying), gathering information, reworking the outline (reclassifying), gathering further information, and back and forth until a final form is achieved.

The very purpose of scientific inquiry itself is not simply finding unique instances of things, but, as philosopher of science Max Wartofsky states (1968), discovering "relations or patterns among the facts, to order them or to link them to each other in some intelligible way" (p. 129). Our chief tool for achieving this is classifying. The process, according to Harvey (1969), may be considered the "beginning point or the culmination of scientific investigation" (p. 326). And while, as philosopher of science Harold Larrabee says (1945), "we are born into a world where most of the familiar things are already pretty thoroughly classified" (p. 247), breakthroughs in science often result from the discovery that something we thought belonged to one category actually belonged to another. Apart from the world of aphorisms,

> I like women with a past and men with a future. (Oscar Wilde)
>
> There are two kinds of people: those who kill themselves
> because they are not Ernest Hemingway or Marilyn Monroe,
> and those who kill themselves because they *are*. (anonymous)

original classifications are rare. But communication specialist Rudolph Flesch (1960) reminds us that new classifications will often completely change our attitudes and our thinking. We should watch out for them and add them to our mental repertoire.

In the last analysis, classifying is at the very core of language. It is our chief means of establishing levels of linguistic meaning, since, to borrow Mill's phrase (1950), "by every general name which we introduce, we create a class" (p. 90). As Harvey (1969) reminds us: "If language were restricted to proper names only, communication would be impossible" (p. 324).

THE CONCEPT OF CLASSIFYING

A Brief Historical Note

A taxonomy is "a system by which categories are related to one another by means of class inclusion" (Rosch & Lloyd, 1978, p. 30). It is based on the mode of thinking that we call classifying. That mode of thought, or analysis, was crystallized by Aristotle in his *Organon* and *Metaphysics* as he worked free from the more Eastern, mystical, synthesizing world of the pre-Socratic philosophers and from Plato as a kind of halfway house. And it has come down to us as our most influential way of organizing knowledge. Classification, and its correlate inductive–deductive reasoning, has attracted some of the best minds of the ages—Bacon, Locke, Darwin, Russell—to name but a few.

It was given a considerable boost by the encyclopedic tradition in France and the desire to organize all of human knowledge under one cover. The need for classifying increased greatly with the birth of modern science. For as Cohen and Nagel (1934) remark, "All sciences in their early days are classificatory" (p. 243). Apart from its use as a tool of science, classification as a legitimate subject of research began only in the 20th century, with Durkheim's publication of "De quelques formes primitives de classifications," in the *Année Sociologique* for 1901–1902.

Induction and Deduction

We will avoid the philosophical debate on inductive and deductive reasoning. The argument has filled volumes. Instead, we will limit ourselves to a few comments on its role in classifying. Psychological studies suggest there is a basic word in a classification hierarchy, an **exemplar**, that corresponds to the level of generic classes in biological taxonomies (Estes, 1994, p. 56). The exemplar is normally a word for which there are distinctive perceptual features; for example, a word such as *dog* rather than *canine* or *animal*. It may also be simply the most common word in the classification set. Superordinate terms tend to lack the perceptual associations of the basic generic term (Miller, 1978, pp. 81–82). Let us keep an eye out for exemplars as we examine the classifications in our corpus.

Hyponymy Relations.

Hyponymy is the relationship of inclusion; that is, of organizing words into taxonomies in treelike diagrams (McCarthy, 1990, p. 19ff). In a hyponymy relationship, the word *diamond* is a **hyponym** of *precious stones*. The term *precious stones* is **superordinate** to *diamond*. Words at the same level as *diamond* (*ruby*, *sapphire*, *emerald*) are **co-hyponyms**, while the entire list of co-hyponyms is called a **lexical**

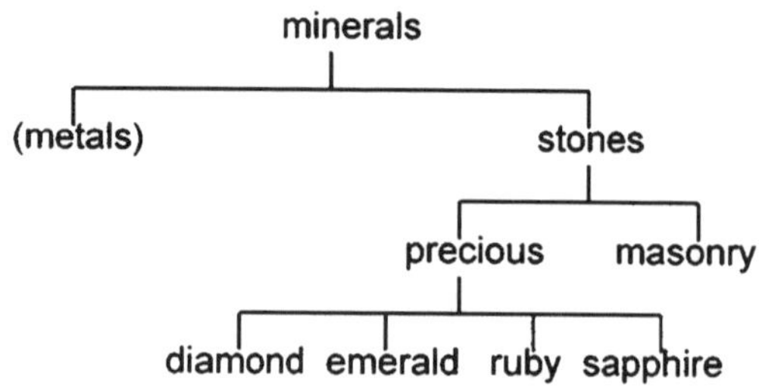

FIGURE 11.1. Hyponymy relations: organizing words into taxonomies in treelike relations.

set. *Precious stones* is a hyponym of *stones*, which is in turn a hyponym of *minerals*, while the broader category of *minerals* would be a **lexical field**. The tree would look like the one shown in Figure 11.1.

Intimate Connections: Observation, Definitions, and Examples

Nothing exists by itself, from the smallest to the largest of things. This is equally true for the skills of thinking. What do we do when we classify? We observe (but not necessarily). We define. We compare. We contrast. We provide examples. We generalize. We measure. How do some of these thinking skills, or modes of perception, bear on the act of classifying? Let us spend a moment on the question.

Observation.

There is no such thing as unbiased observation. As someone once remarked, we see things through our categories. In Wartofsky's words: "In observation, we already classify" (1968, p. 154). Or to quote Nietszche: "Everything that reaches consciousness is utterly and completely adjusted, simplified, schematized, interpreted" (In Medawar, 1969, p. 27). We observe a tree, a boat, a human being—and those various properties of trees and boats and human beings spring up to guide our perceptions. Even the evidence of the senses, prompts Nobel laureate Peter Medawar (1969), "does not enjoy a necessary … authenticity" (p. 41).

Definition.

Classifying and defining are so bound up together that it is hard to imagine one without the other. Indeed, a formal definition *requires* the presence of a class word (Term + Class Word + limiting feature or features) (Darian, 1982). Consider the following examples:

> A mixture is a material containing two or more substances (definition) and can be either heterogeneous or homogeneous (classification). (C50)

In this example, *mixture* is the word being defined, *material* is the class word, and *two or more substances* the limiting feature. In this case, an additional classification is built into the sentence. One of the more common patterns in our corpus is a separate definition, followed by an elaborating classification in the ensuing sentence:

> Those forces that hold oppositely charged ions together or that bind atoms together in molecules are called chemical bonds. The **two principle types** of bonds are the ionic bond and the covalent bond. (C229)

Notice how definition and classification are intertwined in the next example:

> The sum of all chemical reactions that occur within a living organism is **defined** as metabolism. Many hundreds of chemical reactions occur in a typical cell. To make sense of this myriad of reactions, biochemists have **subdivided** metabolism into **two contrasting categories** (classification), anabolism and catabolism. Anabolism is the process by which (definition) ... Catabolism is the process by which ... (definition) (C900)

Examples.

The example, or species, is the only real entity in a classification. The higher levels (taxa) "represent our perceptions of relationships among the groups of species" (Starr, 1991, p. 209). Dogs *per se* do not exist, nor even spaniels, but Cocker Spaniels and Springer Spaniels do. Dogs and spaniels are still abstractions, though admittedly, more tangible than canines. The use of examples is so closely related to classifying that it is sometimes hard to distinguish the example (the species) from the class word. One indication, though not definitive, is the use of the singular for the species. Notice the following use of the word *example*, which in fact indicates a class rather than a species. The actual individual species is the specific cyanobacteria, *Anabaena*:

> There are thousands of diverse eubacteria. Here we will consider a few **examples** ... Cyanobacteria are an **example** ... **Many species** grow as chains of cells that surround themselves with a mucous sheath. Some cyanobacteria, including *Anabaena*, produce ... (B227)

A frequent pattern is a definition, followed by a classification—at one or more levels—terminating in specific examples:

> The phenomenon of two or more compounds having the same number and kinds of atoms is isomerism. There are **two types** of isomerism. In the **first type**, known as structural isomerism ... **For example**, butane and isobutane, ethanol and dimethyl ether ... are structural isomers. (C709)

PATTERNS

As with many of the tools in scientific thinking, the language of classifying results from an interplay of various syntactic, lexical, and discourse elements (Darian, 1982, 1995). In this section, I would like to explore these elements in separate but clearly related treatments.

Syntactic Patterns

Our corpus reveals six syntactic- and discourse-based patterns used to express the idea of classification. I will list them, with examples, and offer a few comments where appropriate. In our notation, the term **classifier** will be used for generic words that signal a classification (e.g., *class*, *group*, or *category* for nouns, or *divide*, *classify*, or *group* for verbs). **CW** indicates a class word (a member of the lexical field) either above (**super**, for superordinate) or below (**hypo**, for hyponym) another class word. The term **subord** (for subordinate) indicates an item at the next lower level in the hierarchy.

1. | **Degree Word** + | **CW(super)** + | **be** + | **CW(hypo)** |
|---|---|---|---|
| **Some** | | | |
| **Most** | | | |
| **All** | | | |
| a) Some of | the proteins | are | enzymes (B49) |
| b) All | cells (except bacteria) | are | eukaryotic (B45) |
| c) About half | of all flowering plants | are | polyploid species (B187) |
| d) Most | monerans & protistans | are | microbes (B222) |
| e) The vast majority | of bacteria | are | heterotrophs of one sort or another (B225) |

Notice that all of the superordinate class words are plurals. As we will see shortly, plurality can be a classification marker. The degree words indicate that there are other items in the lexical set (*co-hyponyms*: closely related words at the same classification level), which are not being discussed in the immediate context. Notice also that the **directionality** of the pattern is from general to specific: The superordinate term appears first.

2. | **CW(hypo)** + | **be** + | **CW(super)** |
|---|---|---|
| a) A carbohydrate | is | a simple sugar or a large molecule composed of sugar units (B28) |
| b) Carbohydrates | are | *the most abundant* biological molecules (B28) |
| c) Enzymes | are | proteins with enormous catalytic power (B61) |
| d) Ribose and deoxyribose | are | in *this category* (B28) |

Directionality is from specific to general: The hyponym appears first. The phrase "the most abundant" (example b) is a quantifier, a concept discussed later in the chapter. Sentences *b* and *c* have a certain "definitional" or descriptive "feel" to them rather than a "classificatory" feel: While there are other biological molecules and other abundant biological molecules, and other proteins with catalytic power, the discussion is not concerned with these distinctions, but rather with the qualities of the specific molecule and enzyme. The last example (d) contains the anaphoric phrase *in this category* and thus relates somewhat differently to its discourse environment.

3. **CW(hypo)**	+ **be**	+ **classifier**	+ **CW (super)**
a) Steel	is	a type	of alloy
b) Maple	is	classed	as a softwood

Directionality is from specific to general.

4. **CW(super)**	+ **classifier(Vpass)**	+ **CW(hypo)**
a) Electrolytes	are classified	as strong or weak, depending on ... (C373)

Though sentences 3b and 4a above both contain a passive verb as classifier, their directionality is opposite. In 3b, *Maple* is a subclass of *softwood.* (The sentence reads from specific to general.) In 4a, *strong or weak* are subclasses of *electrolytes*. (The sentence reads from general to specific.) One could arguably place the "Maple" sentence in Pattern #3 or 4. The reason for including it in Pattern #3 is my sense that directionality is a more powerful organizing device than the classifier (in this case, the passive verb).

5. **CW (super)**	+ **be**	+ **(either)**	+ **CW or CW(hypo)**
a) Covalent bonds	may be	either	polar or nonpolar (B23)
b) Overall, the main metabolic pathways	are		degradative or biosynthetic (B60)
c) Most antigens	are		protein or polysaccharide molecules (B394)
d) We can classify a sample of matter as		either	homogeneous or heterogeneous (B49)
e) A substance	is	either	an element or a compound (C49)
f) A mixture	is		a material containing two or more substances and can be *either* heterogeneous or homogeneous (C50)
g) Properties	are		the personality traits of substances and are classified as *either* physical or chemical (C67)

6. **Sentence** + **enumeration of classes.**
[*Variation in traits is an outcome of certain events that were introduced in earlier chapters.*]

a) These events	fall into	5 broad *categories*: (followed by a list)	(1) … (2) … (3) … (4) … (5) … (B177)
b) There are	3 general categories	of variables: (followed by list)	independent variables … dependent variables … controlled variables (B11)
c) Chemistry	may be broadly *classified*	into two main *branches*:	organic chemistry and inorganic chemistry (C5)
d) Trees	are generally *divided*	into 2 *groups*,	hardwoods and softwoods (Smith 1973, p. 3)
e) Compounds	*fall into*	two general *types*,	molecular and ionic (C56)

In this pattern, syntactic features are subordinate to discourse features. Specifically, the first three slots form an independent sentence by themselves. The series of hyponyms that follows is appended by semi-colons (examples a, b, and c) or by commas (examples *d* and *e*). In examples a and b, the series is presented in the form of a list.

One might arguably place sentences *c*, *d*, and *e* in a class of their own, with the pattern: CW(hypo) + classifier(V) + classifier(NP). One could also include sentence a, although the CW here is anaphoric. It is a matter of classification. As science educator Peter Gega (1994) reminds us, classifying is done to fit a purpose. "What works to fulfill the intent of the classifier is what counts. Objects can be classified in many ways" (p. 73). If our purposes are purely linguistic, we might establish a separate pattern. If they are pedagogical, this more inclusive discourse-based arrangement might be preferable since it reduces our list of patterns by one.

There are four feature differences in our six patterns. The primary feature is directionality (↓) or (↑). Four of the six patterns go from general to specific (superordinate to hyponym). The other three—let's call them secondary features—are the existence of (2) a **degree word**, (3) a **classifier** (adjectival or verbal), or (4) the pattern **(either)** … **or**. Our own classification encounters some difficulty with the secondary features.

For example, Patterns #1 and 5 have the same directionality. Pattern #5 is differentiated from #1 by the existence of (either) … or. However, it also contains two sentences with degree words. How to resolve the overlap? There are two solutions. First is to accept the overlap—some scientific classifications do overlap at certain levels (see, for instance, Figure 11.8). The other is to decide whether one secondary feature is more significant than the other. In the case of Pattern #5, the terms *overall* and *most* acknowledge that there are two major hyponyms in the class, and that other pathways and antigens are not significant.

Lexical Considerations

The lexical elements in classifying can be arranged in four groups:

- nominal classifiers,
- verbal classifiers,
- multiple classifiers and concordances, and
- words in multiple forms.

Nominal Classifiers.

Our corpus reveals 15 or 16 nouns or noun phrases used to indicate a classification. The list includes: *branches*, *category*, *class*, *classification*, *example*, *division*, *families*, *a form of*, *group*, *grouping*, *kind(s) of*, *set*, *sort of*, *subdivision*, and *type of.* Of these, the most common words are: **category, class**, and **type of.**

Verbal Classifiers.

We also find eight or nine verbal forms used to indicate a classification. They include *class*, *classify*, *be classified*, *distinguish*, *divide*, *fall into*, *grouped*, and *separated*. Of these, the most frequent by far is the passive **be classified** (Pattern #4), followed by the verb *fall into* (as in "They fall into three categories").

Multiple Classifiers and Concordances.

The most common concordances between our nominal and verbal classifiers are the following:

classified (pass) into	branches
	groupings
fall into	types
	categories
divided into	groups

Words in Multiple Forms.

The following words appear as different parts of speech:

class (n), class (v), classification
division, divide
group (n), grouping, group (v)

Other Discourse Features

In this section, I would like to discuss three additional features of classifying as they appear in our texts:

1. Relaters
2. Plurals
3. "The Etcetera Factor"

Relaters.

Relaters indicate a special way that different items at the same level (co-hyponyms) relate to each other. Though several items may be at the same level in a hierarchy, there may be an unequal relationship between them—some type of sequence or priority relationship; for example, one of time, cause and effect, and so forth. These relationships are not indicated in classification trees since such graphic devices are schematic rather than elaborative.

Our first example deals with the structure and function of neurons. The text lists their various tasks, then explains that there are

> different *classes* of neurons, called sensory neurons, interneurons, and motor neurons … We can define each class in terms of its role in a control scheme … by which the nervous system monitors and responds to change. The scheme has *receptors, integrators,* and *effector components* …
>
> Sensory neurons are receptors that can detect specific stimuli, such as light energy. *They relay signals to the spinal cord—the integrators* … In the brain and spinal cord are interneurons, which integrate information arriving on sensory lines and then *influence other neurons in return.* Motor neurons *relay information away from the integrators* to muscle cells or gland cells … (B427)

A classification tree of the information would look like the one shown in Figure 11.2. While the three neurons belong to the same lexical set, their relationship is clearly

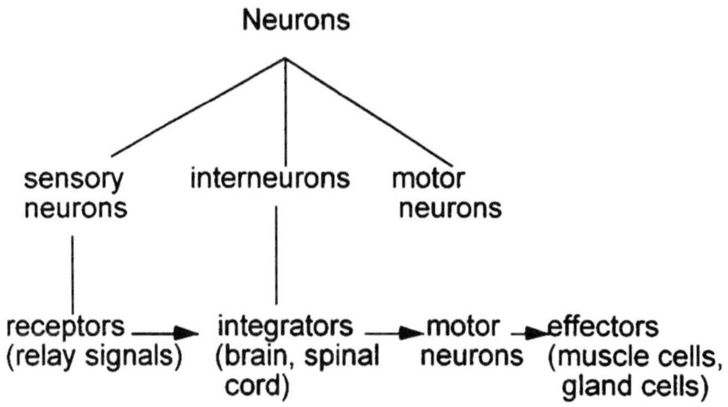

FIGURE 11.2. Classification tree illustrating the structure and function of neurons.

sequential. Impulses are relayed from sensory neurons → to interneurons → to motor neurons. Presumably for this reason, the text represents the relationship as a sequential diagram rather than a classification tree. See Figure 11.3.

Our second example is drawn from information on the immune system. The text, in the form of a sentence list, describes the six kinds of white blood cells responsible for immune responses. This would yield a classification tree with the concept White Blood Cells Responsible for Immune Response as superordinate, and the six specific kinds of cells as co-hyponyms. The sentence list is followed by this passage:

> The cells just listed belong to **two** fighting **branches** of the immune system. Both are called into action during most battles. T cells **dominate** one branch; they carry out a "cell-mediated" response. B cells **dominate** the other branch; they carry out an "antibody-mediated" response. (B393)

The implication here is that there are several other cell-mediated and antibody-mediated kinds of responses and that T cells and B cells *are the most significant.* Indeed, the chapter mentions some of the others later in the text. A tree diagram of this and related information from the text would look like the one shown in Figure 11.4.

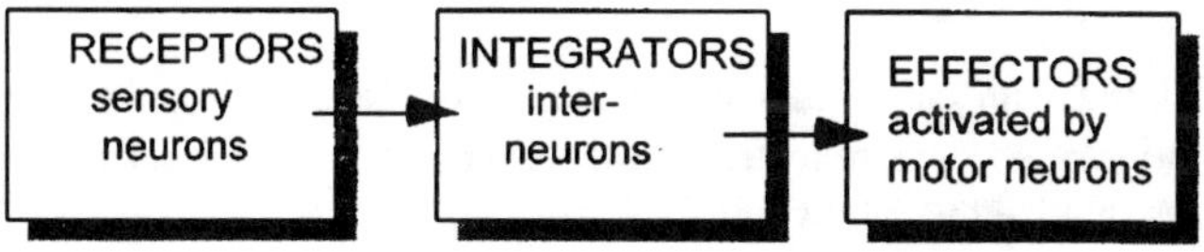

FIGURE 11.3. The same information in diagram form.

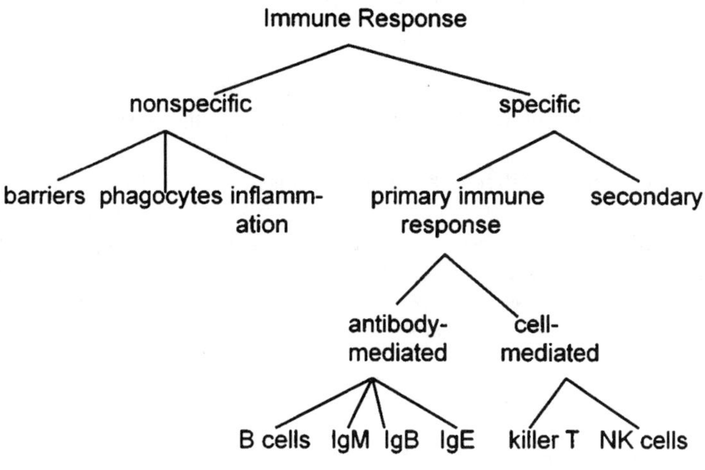

FIGURE 11.4. Classification tree showing elements of the immune system.

Our third example comes from the same source. Notice in the tree diagram above, the level of primary and secondary immune responses. While these are co-hyponyms, they are of unequal power. For the text goes on to state:

> A secondary immune response to a previously encountered antigen *occurs more rapidly* than a primary response, and it *is greater and of longer duration* (B396).

Quantifiers.

Relaters shade over into **quantifiers**, a subclass of relaters that indicates a quantitative difference between members of a lexical set. Our second and third examples above could probably fit this category, inasmuch as they state a greater magnitude of certain hyponyms over others. Many of the degree words in Pattern #1 are further examples of quantifiers:

> *Most* antigens are protein or polysaccharides. (B394)
> *The vast majority* of monerans and protistans are microbes. (B225)
> *All* cells (except bacteria) are eukaryotic. (B45)
> *About half* of all flowering plants are polyploid species. (B187)
> There are 280,000 species of plants … *Most* are vascular plants. (B245)

Plurals.

When dealing with classification statements, the use of a plural in a word that might be read as a species indicates that the example is a member of a class rather than a species. Notice the following:

> Bryophytes are nonvascular land plants … Bryophytes include *mosses*, *liverworts*, and *hornworts*. (B248)

The plurals indicate that there are different species of mosses, etc.; that is, that mosses are not a species.

> *Enzymes* are *proteins* with enormous catalytic power. (B61)
> *Sugars* and other *alcohols* have hydroxyl groups attached to the backbone … (B27)

In other words, there are different kinds of proteins, sugars, and alcohols:

> Different animal viruses infect invertebrates and vertebrates. Among them are the *Herpes viruses* and the *viruses* that cause recurring, worldwide epidemics of influenza. Between 1918 and 1920 alone, a Spanish influenza virus killed more than 20 million people. *Influenza viruses* infect tissues of the upper respiratory tract … (B225)

So there are several types of Herpes and influenza viruses:

> There are thousands of diverse eubacteria … Cyanobacteria **are** (plural) an example …

> **Some** cyanobacteria, including Anabacteria, produce a nitrogen-fixing enzyme ... **Many species** grow as chains of cells that ... (B227)

The plural suggests that *cyanobacteria* is not a species. This is reinforced by the following markers: *some* and *many species*.

The Etcetera Factor.

What I have called the etcetera factor refers to various words that indicate there are other unstated members of a class. These words can be **overt**; that is, specifically stated (*other*, *some other*), or **covert**; that is, implied (*three kinds of* X, suggesting there are other kinds of X). Some quantifiers also have a similar effect ("Most antigens are ... " "About half of all flowering plants ... "). Here are some examples of "etcetera words" or phrases. I have divided them into the following classes: **Other**, **Enumerators**, **Quantifiers**, **Tacit Classes**, and **Miscellaneous**.

Other. An overt marker:

> Chloroplasts of mature leaves contain chlorophylls, cartenoids ... and **other** pigments, each of which absorbs certain wavelengths of light (B73) [Meaning: There are other kinds of pigments.]
>
> Sugars and **other** alcohols ... (B27)
>
> ... fungi and **other** decomposers ... (B241)
>
> As we saw ... crossing over and recombination put new mixes of alleles (variant genes) in chromosomes. We also saw that the structure and number of chromosomes can change through nondisjunction or **some other** abnormal event. (B155)

Enumerators. Covert:

> **Three common** lipids (the glycerides, phospholipids, and waxes) have fatty acids, stretched out like tails. (B29) [Implication: There are other lipids as well.]
>
> **Three kinds of** nucleotides ... are adenosine phosphates, the nucleotide enzymes, and the nucleic acids. (B34; B227)
>
> When that happens, **three types** of white blood cells ... make precise counterattacks. (B393)
>
> There are **thousands** of diverse eubacteria. Here we will consider a few examples ... (B227)
>
> Let's now look briefly at **two** anaerobic routes. Both are fermentation pathways. (B87)

Note the absence of the definite article in sentence five (*two*, not *the two*). This implies there are more than two anaerobic routes.

Quantifiers.

> Overall, the **main** metabolic pathways are degradative or biosynthetic. In degradatative pathways, carbohydrates, lipids, and proteins are broken down in stepwise reactions … (B60)
>
> The **main** degradative pathways are aerobic respiration … and fermentation. (B80)

Tacit classes. When only one class is mentioned, there is a tacit assumption that other classes also exist. For example:

> Four classes of psychoactive drugs are problems in our society. (B439) [Tacit class: psychoactive drugs that are *not* problems in our society]

Miscellaneous.

> Bacteria are the **smallest** cells. (B45) [Implication: There are other larger cells.]
>
> There are **several** different classes of antibodies (Ig) … that enlist the aid of different immune cells or chemical weapons **[Overt].** When bound to antigens, **for example**, IgM and IgG antibodies enlist the aid of macrophages … and IgE antibodies call histerame secreting cells into action. (395) **[Covert]**

For example here suggests there are other Igs as well and is thus being used as an implied etcetera marker. Here is another case in which the word *example* functions as an etcetera marker:

> Photosynthesis and aerobic respiration are **examples** of metabolic activity, which occurs only in living things. (B5) [Implication: There are other kinds of metabolic activity.]

Our final example is from a classification table (B35) with the column headings: *Category*, *Main Subcategories*, and *Examples*. The phrase *main subcategories* suggests there are other subcategories omitted from the table.

Contiguous, Extended, and Nested Classifications

Contiguous and Extended Classifications.

As we have seen, many classification statements are unmarked or implicit. For this reason, classifying may seem less significant in written discourse than are other modes of organization. This is not the case in our science texts, which draw heavily on classifying. In addition to a variety of patterns, we find passages in which a series of classifying statements follow one upon another. Notice the following example:

> … proteins (level 2) are the most diverse of all biological molecules (level 1). They include *enzymes*, which … they also include molecules concerned with cell movements, storage, and transport. Many hormones are proteins; so are antibodies (level 3). (B31)

The passage above includes three classifying statements at three levels of classification. In this way, classifying can be seen more as a discourse-level than a syntactic-level phenomenon. While there are discrete sentence-level patterns, the process of classifying often involves a passage of several sentences, for the writer to present the various levels of the classification. Here, for instance, is a passage containing six levels of classification:

> Four classes of (2) psychoactive (1) drugs are (3) problems in our society (tacit class: psychoactive drugs that are *not* problems in our society). They are (4) stimulants, depressants and hypnotics, narcotic analgesics, hallucinogens and psychedelics ... Stimulants include (5) caffeine, nicotine, amphetamines, and cocaine ... (6) Coffee, tea, chocolate, and many soft drinks contain caffeine. (B439)

A classification tree would look like the one shown in Figure 11.5.

Nested Classifications.

We also find sentences containing **nested classifications**: single sentences containing classifications at more than two levels. For example:

> Many hormones, including sex hormones, are steroids. (B31)

In this sentence, *steroids* is superordinate to *hormones*, which is superordinate to *sex hormones*. The following passage also contains a sentence with three levels of classifying, plus a fourth level in the following (contiguous) sentence:

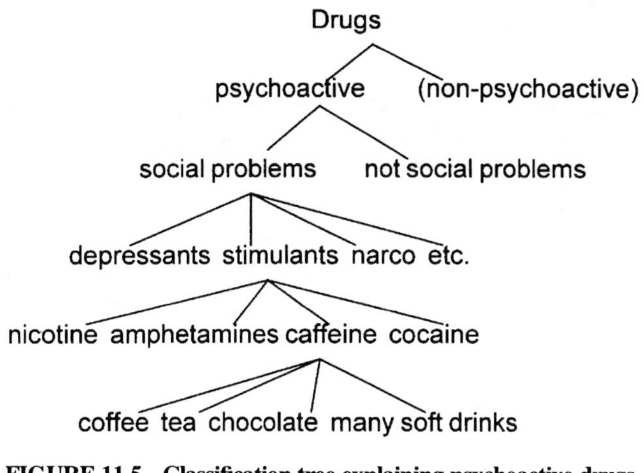

FIGURE 11.5. Classification tree explaining psychoactive drugs.

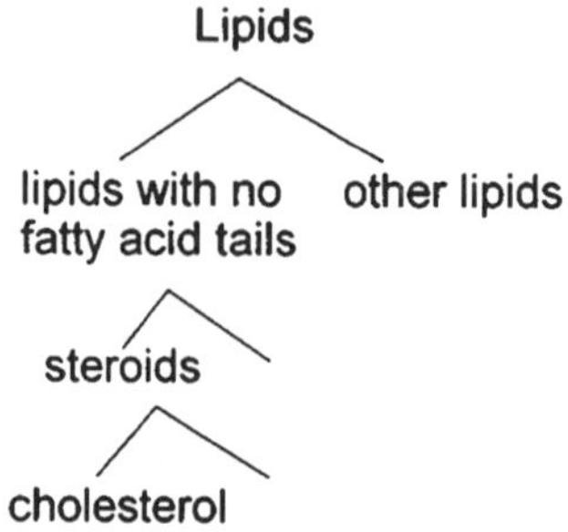

FIGURE 11.6. Classification tree of lipids.

> Among (1) the lipids (2) that have no fatty acid tails, we find (3) the steroids ... You have probably heard of (4) cholesterol. This steroid is a key component of animal cell membranes. (B29)

The classification tree would look like the tree shown in Figure 11.6.

Graphical Representations

In addition to text, there are several graphical formats in our corpus that are used to represent classifications. They include the following:

1. Classification trees and flowcharts
2. Diagrams
3. Line drawings
4. Sentence lists
5. Pictures/Photos
6. Classification tables
7. Charts

The most common device is the classification table (27 instances), followed by photographs (18 instances), classification trees and flowcharts (17 instances), and sentence lists (13 instances). Here are a few observations on these items.

Pictures.

Pictures may seem an unlikely way to represent a classification. But judging from their frequency, they are quite popular. Pictures function in three different ways:

1. the superordinate concept is discussed or mentioned in the text but *not* the species or item, which appears only in the photograph (B244, 246, 275, 287; C31, 208). For example:

Figure 18.5 illustrates some of the diversity that exists among the club fungus. (B243) [SD: This is followed by photographs of three different species.]

Spiders generally are eight-legged many-eyed predators (Figure 19.26). (B274) [Note: followed by photos of three different species. Note also that there is a numerical but no verbal reference to the pictures.]

2. both superordinate and hyponyms are discussed in the text, and several hyponyms are illustrated in pictures (B252, 275). For example:

   Seedless vascular plants once dominated the land (Figure 18.14). Existing members of this group include the lycophytes, horsetails, and ferns. (B249) [SD: followed by photos of the three existing species plus an historical reconstruction.]

3. superordinate and hyponyms are discussed in the text, but at considerable length, sometimes over several pages (B218, 288; C208). In this case, the pictures of specific species serve as *visual summaries*.

Classification Tables.

The tables in our corpus contain between two and five levels of classification in addition to a superordinate term in text or heading. About half the tables do *not* include specific species or examples. This is true even for one that contains five levels of classification (B251). In addition to classifications, some tables also offer other kinds of information.

Flowcharts and Classification Trees.

The use of flowcharts and classification trees differs markedly in our two volumes. The chemistry text uses flowcharts four-to-one, with all of them flowing downward ↓. The biology text, on the other hand, with its emphasis on evolutionary processes, makes greater use of trees (six compared to one), which are all directed upward ↑. It has only one flowchart, used to describe the branches of the immune system. The reasons for the difference are not apparent. But since biology is more time-oriented than chemistry (evolution, physical development) and since language is partly metaphorical (Lakoff and Johnson, 1980), one can logically conceive the process of evolution and growth in an upward direction, especially since, as Lakoff and Johnson point out, *up* connotes positive change, as does evolution.

Sentence Lists.

The sentence list is a halfway house between text and visuals and is the fourth most common graphic.

PROBLEMS IN CLASSIFYING

Directionality

As we have seen in our study of patterns, a potential problem in the reading (or writing) of classifications is *directionality*: It may not always be clear whether the direction of a classification statement is general-to-specific or specific-to-general. Patterns #1 and 5 are general-to-specific. Patterns #2 and 3 are specific-to-general. And Pattern #4 contains an example of each.

Division and Classification

Along with numerous classification statements, our corpus contains many instances of composition (parts, partition, division) describing the relation of parts to wholes. It is sometimes hard to distinguish the two concepts. We can normally identify division by key words—nouns such as *parts*, *components*, *section* or *subsection*; or verbs such as *composed of*, *consists of*, *constructed of*, *contains*. However, we find cases where some of these partition words are used for classifying:

> Even the most complex animal **is constructed of** only **four kinds of** tissues, called epithelial, connective, muscle, and nervous tissues ... Nearly all of the cells in these body tissues are collectively called somatic cells ... The exceptions are the germ cells, the only cells in the animal body that give rise to sperm and eggs. (B337)

The verb *constructed of* is normally considered a division marker, but here it is followed by the phrase *kinds of*, which usually indicates a classification. A clue to the status of the word *tissues* is its plurality, indicating it is above the species, or example level. This is born out by photos on the following page, showing different kinds of epithelials, etc.

The next example further illustrates the connection between the two modes. The classifying word *classes* actually *is* a part of the superordinate *nucleic acids*:

> Nucleic acids are complex chemicals that combine several different *classes* of smaller molecules (definition) ... We will start the examination of nucleic acids by learning about a critical **part** of these molecules, two *classes* of heterocyclic bases, the purines and the pyramidines.
>
> There are five major bases commonly found in nucleic acids—two purines (adsenine and guanine) and three pyramidines (cytosine, thymine, and uracil). (C843)
>
> A DNA unit **is composed of** *four kinds of* nucleotides, the *subunits* of nucleic acids. A nucleotide **consists of** a 5-carbon sugar, a phosphate group, and one of the following nitrogen-containing bases: adenine, guanine, thymine, cytosine. (B145)

Ultimately, division is a relationship of parts-to-whole while classification is one of class-to-subclass. The characteristic clause type in division is possessive, with the verb *have* ("All vascular plants have well-developed roots, stems, and leaves"). In terms of graphical representations, division tends to rely more on diagrams and labeled pictures (Halliday & Martin, 1993, pp. 174–175).

Overlap and Inverted Sequences

Classifying is an attempt to make sense of the seemingly infinite number of things in the universe—to *find* or *create* order. In both cases, a degree of arbitrariness creeps into the process. For in the last analysis, classifications are not things at all (Ellen, 1979, p. 27), but our efforts—based on memory, reasoning, and fallible perceptions—to organize that universe. In this we are influenced by our Western passion for symmetry, which is encouraged by mathematicians and fathered by Aristotle. Unfortunately, such patterns do not necessarily correspond to the natural world. Thus, it is reasonable to expect inconsistencies in the classifications from our corpus.

Accordingly, we find cases in our texts, of overlapping categories as well as inverted sequences in which a hyponym appears above its superordinate. A few examples follow. The first illustrates the problem of ambiguity, or overlap, in which a hyponym may belong to more than one superordinate (category). The example comes from a chapter on immunity, which describes the body's various kinds of defense responses, some of which are nonspecific (occurring when any kind of invasion is detected) and others specific (the mobilization of white blood cells to specific kinds of invaders). The text goes on to describe various kinds of nonspecific defense responses, as shown in the classification tree (mine) in Figure 11.7.

A following section on the Complement System states:

> When certain bacterial ... cells invade a tissue, about 20 plasma proteins interact as a system—the complement system—*with roles in both nonspecific and specific responses.* (B391)

Thus, the anomaly (overlap) of the complement system is a partly specific, partly nonspecific response. A second example deals with metabolism and the concept of *metabolic pathways* (L1), which are defined as:

> an orderly series of reactions, the steps of which are quickened with the help of specif-

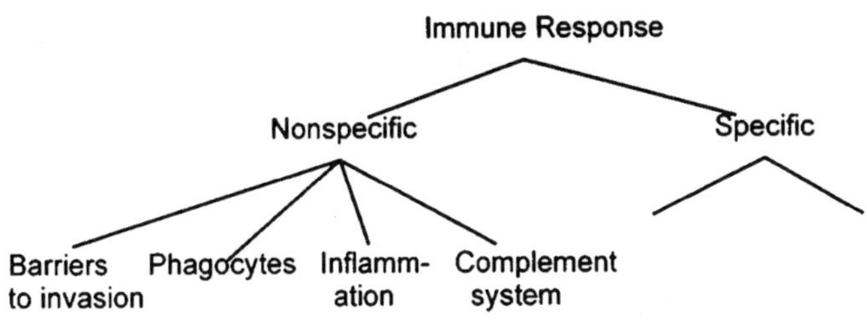

FIGURE 11.7. Classification tree showing various types of nonspecific immune responses.

ic enzymes. **Most** sequences are (L3) *linear*; some are *cyclic*. Branches often link different pathways, with products of one pathway serving as reactants for others. Overall, the main metabolic pathways are (L2) *degradative (catabolic)* **or** *biosynthetic (anabolic)*. (B60)

The highest superordinate (*metabolic pathways:* Level 1) is followed by *linear* and *cyclic sequences* (Level 3), which, as the following tree illustrates, occur *below* the actual names of the main metabolic pathways (Level 2). The tree includes the terms *aerobic* and *anaerobic*, which are mentioned in a previous passage. The logical sequence is like that shown in Figure 11.8.

Our last example illustrates an inverted sequence (Figure 11.9), in which the hyponym appears above its superordinate (C326):

(Level 1) Most of the *substances* we encounter in our daily lives are **(L2)** *mixtures*. Often they are **(L3)** *homogeneous mixtures*, which are called *solutions*. When you think

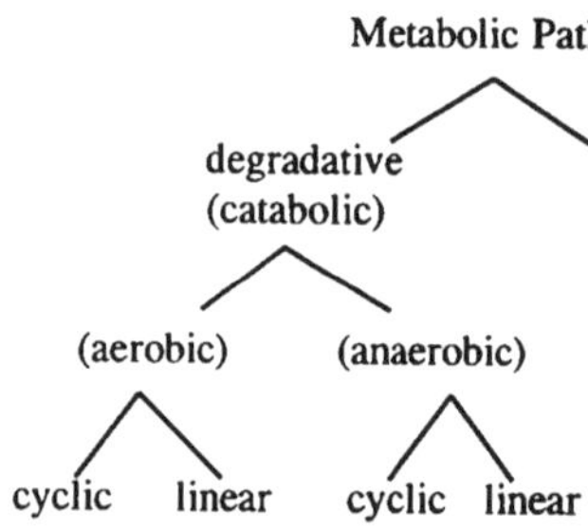

FIGURE 11.8. Classification tree illustrating various types of metabolic pathways.

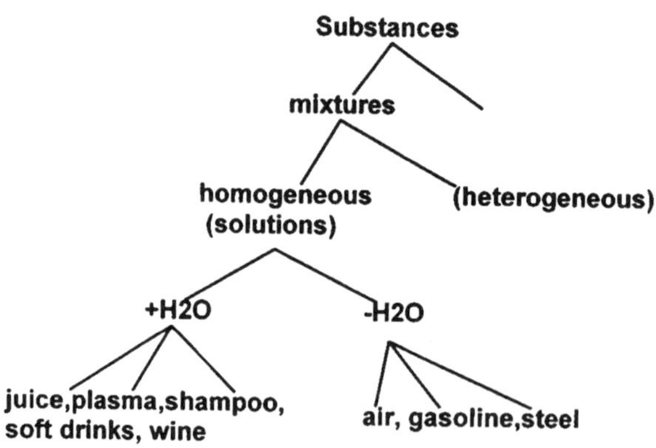

FIGURE 11.9. Classification tree revealing an inverted sequence in the text.

of a solution, **(L5)** juices, blood plasma, shampoo, soft drinks, or wine come to mind. These solutions all have **(L4)** *water* as a main component. However, many common items, such as **(L5)** air, gasoline, and steel are solutions **(L4)** *that do not contain water.* (C326)

CONCLUSIONS

Several theoretical issues were raised at the beginning of this chapter. Let us examine them in terms of our corpus and the examples chosen for illustration and analysis. An initial matter is the actual function or functions of classifying. Bruner et al. (1956) and Panchen (1992) suggest that one of the *functions* of classifying is to reduce the overwhelming number of items that must be processed by creatures of limited capacities, humans and otherwise. One of our excerpts on the immune system indicates that this function operates at the microscopic level as well as when the immune mechanism classifies an invader as either specific or nonspecific (Fig. 11.4) and chooses a form of defense accordingly.

Crowson (1970) and Harvey (1969) consider classifying as our primary thought process for understanding, or finding order in, the world; Wartovsky (1968), as a means of discovering relations or patterns among facts. The matter of order, or intelligibility, of course, may be culture-bound or culture-independent (Bergenholtz & Tarp, 1955). Recalling our classification tree of minerals and stones, we can logically imagine a culture where diamonds are *not* considered precious stones. Or take the excerpt on psychoactive drugs that cause social problems in our society. These include narcotics and caffeine products such as coffee, chocolate, and many soft drinks. In certain traditional societies, narcotics were not a social problem and even formed part of accepted religious ritual. The author of our biology text is careful to stipulate, however, that the classification applies to "our society" (B439).

A final example is from the text and classification tree explaining the term *substance*. To the nonspecialist, items like blood plasma, shampoo, air, and steel would not be perceived as related. To the scientist, they are all examples of homogeneous solutions. Can we say that the gulf between specialist and nonspecialist from the same culture is a culture-bound difference? To a certain extent, the answer must be yes.

As for Estes' thesis (1994) that classification is basic to all of our intellectual activities, we have already observed its interactions with other thought patterns in our corpus. An obligatory part of *formal definitions*, for example, is a class word (*Enzymes are proteins ... Many hormones are steroids ...*). In addition, classifying is one of the chief means of *expanding* a definition (*The forces that ... bind atoms together in molecules are called chemical bonds. The two principle types of bonds are the ionic bond and the covalent bond*).

We have also seen how the process of *exemplification* flows naturally from the act of classifying. Examples provide the ultimate "anchor"—the specific species or "instantiation"—for the superordinates of the lexical field: endpoints in the semantic journey from immune responses to Killer T cells, from drugs to coffee, from lipids to cholesterol.

Generalization also plays its part here, as one of the most basic processes of thinking in general and scientific thinking in particular. Hypotheses are generalizations. The uppermost term in the lexical field is a generalization. Minerals, proteins, drugs, tissues, as we have seen, are concepts, or ideas, and not things. As such, they are built on one or more levels of hyponyms beneath them. We can never know for sure whether an early cave dweller called that big orange cat with black stripes a tiger or an animal. From the standpoint of survival, he would have been safer calling it a tiger. But whichever it was, the process of classifying was central to his perception.

Another approach to the operation of classifying is the concept of *exemplars*—words that correspond to the level of generic classes in biological taxonomies. Estes (1994) and Miller (1978) suggest that exemplars are words that normally have distinctive perceptual features or are possibly the most common word in the classification set. One problem with the perceptual nature of exemplars is that, in fields such as biology, chemistry, and especially physics, an exemplar might be microscopic or submicroscopic and so does not *have* distinct perceptual features. Quite the contrary, several exemplars from our corpus appear at the highest level of abstraction (the lexical field), including *metabolism* (p. 185), *neurons* (p. 190), and *virus* (p. 192).

On the other hand, the term *cholesterol*, in the discussion of lipids (p. 18), probably qualifies as the exemplar because of its considerable popularity. A final example is the explanation of the term *substance* (p. 200), which lacks both perceptual qualities and widespread use. And we have to descend three further levels before we reach the most common words of the set (*juice*, *wine*, *gasoline*). In this case, these common everyday words could be part of a variety of lexical fields and so do not especially exemplify substances.

Finally, let us touch on geographer David Harvey's assertion (1969) that classifying may be considered the beginning point or culmination of scientific investigation. The three variables in the sentence are *beginning point, culmination*, and *scientific investigation*. Unfortunately, scientists often dislike the phrase *scientific method*, perhaps because it suggests a rigid sequence of steps, whereas certain steps in scientific investigation tend to be sequential while others are not (Darian, 1995, 1996). Hypotheses, for example, may result from experiment, observation, or an internal thought process. Let us take as a working definition of scientific investigation: the systematic examination natural phenomena. What support for Harvey's claim do we find in our corpus?

From the standpoint of "top-down" analysis (↑ ↓), each step down the ladder leads to a more specific item and ultimately to the smallest working part or process in the organism, if we are dealing, say, with biological phenomena. Apart from applications, in medicine, for example, this might be considered the culmination of the scientific investigation, to use Harvey's phrase. However, the terms *beginning point* and *culmination* are not necessarily synonymous with the highest (↑) and lowest point (↓) of a taxonomy (lexical field and hyponym). In our botanical excerpt (p. 197), people began to study ferns before they classified them as seedless vascular plants. In this case, the beginning point is at the species (hyponym) level and the culmination is the lexical field at the "top" of the taxonomy. By contrast, the tree diagram of the immune response (Figure 11.4), in addition to studies in the history of science (Conant, 1950–1954), suggests that the concept of immunity appeared long before knowledge of its specific responses (B cells, etc.).

We might learn a great deal more about the starting point (↑ ↓) of classifications by comparing the anthropological work on folk taxonomies with those in formal branches of science, but that would take us too far afield from our study.

Exercises and the Teaching of Classification

Trimble (1985) suggests three kinds of classifications: **complete**, **partial**, and **implicit**. A complete classification contains three kinds of information: a) the name of the class (the set), b) members of the class (the subset, or hyponyms), and c) the **basis** for classifying (how members are similar and different from each other). A partial classification includes the members and name of the class but *not* the basis, while an implicit classification has all the classifying elements, but they are not labeled as such (Trimble, 1985, p. 80ff).

According to this scheme, the best place to start in teaching classifications is recognition of complete classifications, beginning with two levels (one superodinate and one hyponym). This means designing activities to help students recognize and define the class word and the members and be able to state the basis for differentiating *and* relating them. Trimble warns that unless students grasp the concepts stated in complete classifications, partial classifications will be hard to understand, and implicit versions almost impossible. We should keep in mind that this framework involves discourse-level descriptions. As such, it is important that the instructor work with discourse-length material, at least for comprehension purposes.

A more robust or canonical form of classification would contain additional features: items we have seen in examples throughout the chapter. Take this passage:

> (1) Sailboats (2) may be divided into (3) two (4) categories: (5) planing boats and displacement boats. (6) A planing boat rests on the surface of the water. A displacement boat has part of its hull (or body) under the water.

We can recognize here six specific classifying elements:

1. the name of the class (sailboats)
2. a verbal classifying term (may be divided)
3. the number of class members mentioned (two)
4. a plural classifying word (categories)
5. two or more members of the class (planing boats and displacement boats)
6. the basis for classification (on the surface versus under the water)

In teaching classification, one should emphasize those classifying words—nominal and verbal—that help students recognize and understand the discourse pattern.

Implicit classifications are the hardest to spot since, by definition, they lack the lexical markers that help identify them. One way of helping students extract the information is presenting a passage in incomplete mode, followed by the items you would like them to identify (Trimble, 1985). A handout would look something like this:

> Three general categories of **softwood lumber** are cut, depending on their eventual uses. They are yard lumber, factory lumber, and structural lumber. Yard lumber is used for light construction such as flooring, roof planks, and siding. It comes from the middle part of the tree and is not especially strong. Shop lumber is used for making items such as doors and cabinets. It comes from the outer part of the log and has medium strength. Structural lumber is used for heavy construction and supporting heavy loads. It comes from the heartwood of the log (near the center) and is the strongest material possible. (Smith, 1973, p.9)

Class:	*Softwood lumber*
Members of the class:	*Yard lumber, shop lumber, structural lumber*
Basis for Classification:	*Strength, part of the tree*

Here are some other ways to help students understand the complexities of classification discourse in general. They draw on two of the visual modes used by our texts to express them: trees and tables. Have students take a passage—complete, partial, or implicit—and outline it in the form of a classification tree. Our recent passage would look like the one in Figure 11.10.

Or you might include a skeleton classification table and have students fill it in. As we saw earlier in our analysis, other information is sometimes included along with classification. In this case, the table might look like Figure 11.11.

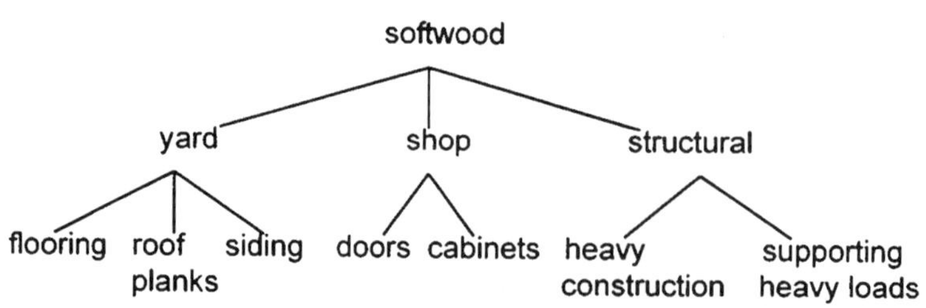

FIGURE 11.10. Classification tree used as an outline.

SOFTWOOD	USE	ORIGIN	STRENGTH
Yard	*light construction*	*middle of tree*	*low*
Shop	*doors cabinets*	*outer part of tree*	*medium*
Structural	*heavy construction*	*center of tree*	*high*

FIGURE 11.11. The same information in table form.

Here are a few final suggestions for teaching classifications:

1. *Chapter Outlines*. Have students do a classification tree for those chapters that lend themselves to it.
2. *Start in the middle* and have students go in both directions: (↑) and (↓). Push it to the limit. For example, have students start with their country or city and go by steps to *the universe* and to their own home. *Levels*. Have them start with two levels, then increase to three, then four, or maybe five.
3. *Definitions*. As we have seen, classifying and defining are closely related. When teaching formal definitions, point out the interaction between the two modes. Apparently, it is very easy for students to grasp their relationship to classification statements (Trimble, 1985).
4. *Games*. There are a thousand things in the mind and in the world waiting to be classified—and a thousand ways of classifying them. The game 20 Questions can help students develop classifying questions (both broad and narrow). How many ways can you think of to classify the capital letters in the Roman alphabet? (There are literally thousands.) Have students classify familiar things orally or in writing: things in class, things they are interested in.

Above all, remember: classifying can be fun. It is a true intellectual challenge, at any age. And to make it less than that is to turn the palace of the intellect into a dusty attic. May we never do this to our students. Or to ourselves.

REFERENCES

Bergenholtz, H., & Tarp, S. (1995). *Manual of specialized lexicography*. Amsterdam and Philadelphia: John Benjamins.

Bruner, J., Goodnow, J. J., & Austin, G. A. (1956). *A study of thinking*. New York: Wiley.

Cohen, M., & Nagel, E. (1934). *An introduction to logic and scientific method*. New York: Harcourt Brace.

Conant, J. B. (1950–1954). *Harvard case histories in experimental science* (Vols. 1–2). Cambridge, MA: Harvard University Press.

Crowson, R. (1970). *Classification and biology*. New York: Atherton.

Cruse, D. A. (1986). *Lexical semantics*. Cambridge: Cambridge University Press.

Darian, S. (1982). The role of definitions in scientific and technical writing. In J. Hoedt (Ed.), *Pragmatics and LSP* (pp. 27–47). Copenhagen, Denmark: Copenhagen School of Economics/UNESCO.

Darian, S. (1995). Hypotheses in introductory science texts. *IRAL. 33,* 84-108.

Darian, S. (1996). The language of experiments in introductory science texts. *Fachsprache* (in press).

Ellen, R. F. (1979). *Classifications in their social context*. New York: Academic.

Estes, W. (1994). *Classification and cognition*. New York: Oxford University Press.

Flesch, R. (1960). *How to write, speak, and think more effectively*. New York: New American Library.

Gega, P. (1994). *Science in elementary education*. New York: Macmillan.

Halliday, M. A. K., & Martin, J. R. (1993). *Writing science*. Pittsburgh, PA: University of Pittsburgh.

Harvey, D. (1969). *Explanation in geography*. New York: St. Martin's.

Hein, M., Best, L. R., Pattison, S., & Arena, S. (1993). *College chemistry.* Pacific Grove, CA: Brooks Cole.

Lakoff, G., & Johnson, M. (1980). *Metaphors we live by*. Chicago: University of Chicago Press.

Larrabee, H. (1945). *Reliable knowledge*. Boston: Houghton Mifflin.

Lehrer, A. (1974). *Semantic fields and lexical structures*. Amsterdam, Holland: North-Holland.

McCarthy, M. (1990). *Vocabulary*. London: Oxford University Press.

Medawar, P. (1969). *Induction, intuition, and scientific thought*. Philadelphia: American Philosophical Society.

Mill, J. S. (1950). *A system of logic*. New York: Hafner.

Miller, G. A. (1978). Semantic relations among words. In M. Halle, J. Bresnan, & G. A. Miller (Eds.), *Linguistic theory and psychological reality*. Cambridge, MA: MIT Press.

Panchen, A. (1992). *Classification, evolution, and the nature of biology*. Cambridge: Cambridge University Press.

Rosch, E., & Lloyd, B. (Eds.). (1978). *Cognition and categorization*. Hillsdale, NJ: Erlbaum.

Smith, R. C. (1973). *Materials of construction*. New York: McGraw-Hill.

Starr, C. (1991). *Biology: Concepts and applications*. Belmont, CA: Wadsworth.

Trimble, L. (1985). *English for science and technology.* Cambridge: Cambridge University Press.

Wartofsky, M. (1968). *Conceptual foundations of scientific thought*. London: Macmillan.

part V

Pedagogical Approaches to Science Writing

12

Changing Moorings in Scientific Writing: Suggestions to Authors, Allusions for Teachers

Philippa Jane Benson
University of Pittsburgh

Many are familiar with Lewis Thomas, Stephen J. Gould, Steven Hawking, and the small handful of other scientists who have been popular and successful as both researchers and as wordsmiths. The reading public seems in awe of these scholars who can illuminate the obscure corners of their expertise with clarity and grace. However, some scientific peers of these authors lack this awe and instead wonder why these researchers spend so much effort publishing beyond their clan, since such vastly different tasks constrain those who write to create new knowledge in the sciences and those who try to help the uninitiated understand this knowledge once it has been amassed and accepted. In the process of writing science, empirical researchers—although bound by the dictum of purposefully recording observations in the course of testing hypotheses—are also on shifting ground.

One of the goals of this chapter is to give readers a sense of the breadth of influences on arrangement and style affecting those writing the natural sciences, using biology as a specific example, and to describe some of the challenges authors face in responding to these differences. Of course, with the diversity of scientific disciplines today, virtually everything said here is sure to be untrue in particular contexts. Nonetheless, I hope readers will forgive the generalities and take them at face value. To this end, I begin by specifically contrasting what I mean by "science" writing and "scientific" writing, and by providing an example of each. I focus on the recent evolution of one particular style and format guide in the sciences, that put out by the

Council of Biological Editors (CBE), the first edition of which was published in 1960 and the most recent 6th edition in 1994. To underscore the changing role of professional style guidebooks in guiding authors, I then review the "Instructions to Authors" from a number of prominent journals and contrast them in the context of the perspectives of those who develop or enforce these discourse codes. To support those descriptions I include sections of text that are aligned to the requirements of particular styles, in part simply to show the contrasts that exist. In light of these discussions, I argue that the traditional approach to the teaching of scientific writing, that based on the Introduction-Method-Results-And-Discussion (IMRAD) format, no longer serves teachers or students well, as that format no longer encompasses the range or writing styles that a writer in the sciences needs to be able to master and wield to succeed in his or her field. An additional goal of this chapter then is to provide new perspectives on teaching scientific writing to the scientists and writing teachers who are increasingly called upon to improve both the basic and the scientific literacy skills of their students.

SCIENCE WRITING AND SCIENTIFIC WRITING

Antoine Lavoisier, the 18th century scientist often considered the father of modern chemistry, is well known not only for his commitment to the enforcement of the scientific method (e.g., the precise use of instruments, analytical rigor, and experimental verification) but also for his part in the debate about the role of language and nomenclature in revising the practice of science. Lavoisier begins his collected works by saying:

> The impossibility of isolating language from science and science from language comes from the fact that every natural science is necessarily made of three things: the series of facts that constitute the science; the ideas that bring these facts to mind; and the words in which the facts are expressed. The word must bring out the idea; the idea must represent the fact: they are three impressions of the same stamp. (Lavoisier, 1965, p. 2)

Lavoisier wrote in the midst of a critical phase in the evolution of the modern scientific method and his emphasis was on the role of the precise use of language, that is the use of terminology and nomenclature, in developing what we now know as the scientific method. At the same time, however, his expression of the unique ability of human language to fuse external fact and internal thought captures the essence of scientific writing as a primary example of when thinking and written expression become purposefully and intimately bound.

A plethora of modern texts are available to help writers learn how to fuse language and thought in such a way that they are successful in publishing in the sciences (e.g., Booth, 1994; Day, 1992; Day, 1994; Ebel, Bliefert, & Russey, 1990; McMillan, 1997;

Pechenik, 1993). However, most of these texts can be considered cookbooks of sorts because the vast majority of their pages are filled with specifics on what information should go into sections of a scientific paper: the abstract, introduction, materials and methods' results, discussion, acknowledgments, and references. They are cookbooks also since they focus on teaching writing and not teaching science; the problem is that writing about science and doing science are not two separate activities.

In general, these texts lay out standard information about what belongs in various sections of a scientific paper and put forth guidelines for good style. Some texts do briefly touch upon the idea that a research paper is an argument founded on supporting evidence and tangentially explore activities crucial for learning science, such as how to read or represent data. However, to give students the impression that getting these recipes under their belt will get them where they need to go is, perhaps, doing them a disservice. The essence of what makes scientific writing good is inherently subjective if one believes that "good" writing is measured solely by readability and style. However, what makes scientific writing good to discipline-specific readers is whether the science that is being argued is relevant and sound, regardless of whether it is in more or less elegant language.

An important place to begin a discussion of these particulars is to distinguish between "science" and "scientific" writing. Although a distinction between these terms may at first seem to be superficial semantics, the distinction is important to make because the two are often unwittingly confounded, particularly by teachers who read about and perhaps on occasion write about—but do not design and conduct—scientific experiments. For the sake of this discussion, we can think of scientific writing as text written by practicing scientists, for persuasive purposes, and for specific audiences who often have some knowledge of the topic at hand. The more "scientific" the writing, the more limited the audience and distribution of the writing and the more the writer expects that readers will be neither confounded nor turned off by information presented in specialized language, or in dry and unamusing detail. The purpose of scientific writing is, plainly, to persuasively argue to a particular community of readers that they should accept the interpretation of new or previously accepted information as sound enough to function as the basis for the inspection or creation of yet additional new knowledge.

Consider, for example, the information presented in the opening paragraph of an article reporting on the mating of the fruit fly. This text was published in the prominent scientific journal *Nature*. The internal citations have been deleted from the following quotation as well as from all other quotations from scientific journals cited throughout this article.

> Female *Drosophila melanogaster* with environmentally or genetically elevated rates of mating die younger than controls. This cost of mating is not attributable to receipt of sperm. We demonstrate here that seminal fluid products from the main cells of the male accessory gland are responsible for the cost of mating in females, and that increasing exposure to these products increases female death rate. Main-cell products are also involved in elevating the rate of female egg-laying, in reducing female receptivity to fur-

> ther matings and in removing or destroying sperm of previous maters. The cost of mating to females may therefore represent a side-effect of evolutionary conflict between males. (Chapman, Liddle, Kalb, Wolfner, & Lartride, 1995, p. 241)

Chapman and her colleagues wrote this introduction in quite clear, concise language. However, though the implications of these findings may intrigue some nonspecialized readers, many would not continue to read on with zeal.

Science writing, on the other hand, is writing about scientific topics for nonexpert audiences for the purpose of informing and entertaining. The intent of this kind of prose is to explain complex information in digestible and even pleasant language. Science writing is for broad, lay audiences who find these texts in nonspecialized publications such as newspapers, popular magazines, and nonfiction books. Consider, for example, the angle spun on Chapman et al.'s research, above, as put forth by the *New York Times*' Natalie Angier (1995). This introduction clearly makes the research substantially more appealing and more immediately understandable.

> Whether in the towering tragedies of Sophocles or the basement pulp of Hollywood, sex and death are intertwined: yin and yang, strophe and antistrophe, head and tail of the same mortal coin. So it is in nature, where the urge to reproduce is fraught with risk, and the cost of raising young drains a parent of resources that might otherwise be devoted to keeping oneself alive just a little bit longer.
>
> Scientists have now discovered yet another reason sex can kill, at least in the case of a favorite laboratory creature, the fruit fly. They have determined that a compound in the male's semen is toxic to the female, and that over the long run (which, by fly standards, means a few weeks) the ingredient significantly shortens her life. And the more partners a female fly consorts with, the sooner will her tiny wings and heart cease to beat. (p. B7)

The point here is to start the meat of this discussion on a common ground of understanding that writing about science for practitioners is a very different business than writing for those who are—and only can be—tangentially invested in either its process or its success. The style in scientific writing is important and can affect the outcome of the to-be-or-not-to-be-published decision that so many editors must render. However, at its heart, editors of scientific publications make their decisions based on the soundness of the science and, as importantly, on their judgment of the relevance of the topic, the findings, and the conclusions to readers. After many years as an active scientist and as a participant in the activities of both the Society for Scholarly Publishing and the Council of Biology Editors, Day (1994) concludes:

> Good scientific writing is not a matter of life and death; it is much more serious than that. The goal of scientific research is publication. Scientists, starting as graduate students, are measured primarily not by their dexterity in laboratory manipulations, not by their innate knowledge of either broad or narrow scientific subjects, and certainly not by their wit or charm; they are measured, and become known (or remain unknown) by their publications. A scientific experiment, no matter how spectacular the results, is not completed until the results are published. In fact, the cornerstone of the philosophy of sci-

> ence is based on the fundamental assumption that original research *must* be published; only thus can new scientific knowledge be authenticated and then added to the existing database that we call scientific knowledge. (p. ix)

SHIFTING GROUND: CHANGES IN THE CBE MANUAL

Over the past decade, however, the playing field has changed significantly for writers in scientific communities. According to the Science Reference and Information Service of the British Library, the number of serial publications in the sciences has grown by nearly 11,000 since the early 1970s (Document Supply Centre, 1996). Major publications, those that everyone in a particular field reads and where everyone (particularly those on tenure tracks) want to publish, are very selective about what they will publish. For example, the journal *Nature* is sent 7,500 manuscripts a year but publishes only about 650, for an overall acceptance rate of about 8.5 percent (Peter Wrobel [general editor of *Nature*], personal communication, November 11, 1996). At the same time, the criteria for publication from one journal to another also widely differ, due to differences in disciplinary focus and traditions as well as to the occasional idiosyncrasies of some editors. Authors looking to publish a manuscript must develop their text in only the most basic forms before selecting which publication they want to submit their work to. A writer who is wed to particular conventions of presenting information, including the order of information, the articulation of purpose, or the style of prose, may hamper his or her ability to get into print. One caveat I must put in here is that what I say here applies more to authors who are not very firmly established in their respective fields. Once a scientist is well established, she or he may be given a considerable amount more leeway in publishing than an author in the process of establishing a reputation.

One way of looking at the evolution of diversity of style and forms of scientific writing is to look at changes in the last two editions of the manual for authors, editors, and publishers put out by the Council of Biology Editors (CBE), aptly entitled "Scientific Style and Format." The first edition of this manual was published in 1960 with subsequent editions in 1964, 1972, 1978, and 1983. These editions changed over the years to accommodate growth in the fields it covered, but no previous revision was as dramatic as that between the 5th and the 6th. The 6th edition of the CBE manual was issued in 1994 and is different in both organization and substance from its predecessors.

The 5th Edition

Like the previous four editions, the 5th edition appears as a compact and elegant manual with a total of 324 pages. All but one of the contributing editors and consultants for the edition were based in U.S. organizations (see list in 5th edition column in Table 12.1). The text begins with a short chapter on ethics, a term which appears neither in the table of contents nor the index of the subsequent 6th edition. The second through sixth chapters (73 pages) discuss how to plan and prepare a manuscript by reviewing various types of scientific communication (e.g., research articles, theses, posters, book reviews, editorials, grant proposals) laying out stages of drafting, providing a mini style guide, and by providing guidelines for preparing references and illustrations. The next eight chapters briefly discuss the editorial review process, copyright issues, procedures for publishing, proofing, and indexing a text, and conventions for using punctuation, as well as writing numbers and symbols (72 pages).

The thirteenth and lengthiest chapter in the book (85 pages) reviews recommended style in special fields, including plant science, microbiology, animal sciences, chemistry and biochemistry, geography, and geology. Specifics here include, for example, guidelines for use of nomenclature and usage in plant and animal sciences, microbiology, chemistry and biochemistry, from how to properly refer to bacteria and viruses to how to write molecular weights. In short, this section of the manual lays out the guts of scientific language, those fine points of specialized discourse that experts in respective fields must be able to both understand and use. The manual winds down with a tiny section on word usage, followed by a list of secondary services for literature searching and an annotated bibliography.

The 6th Edition

The 6th edition of the CBE manual, at 825 pages, is in sheer bulk a much more substantial text than the 5th edition. Unlike all the previous editions, it covers "all scientific disciplines" by reviewing how to write about time systems, the electromagnetic spectrum, all known components of all things living from genes to cells to fungi and lichens to animal life. It documents guidelines for discussing human history and society, the earth, and astronomical objects. Previous editions' recommendations on how to write and submit papers for journal publication have been dropped, and recommendations are confined to very limited stylistic conventions, such as how to properly punctuate, abbreviate, capitalize, and spell.

The text's chapters are grouped into five parts. The first part (18 pages) provides an overview and some justifications for the substantial changes in the manual as well as a primer on the history of scientific notation. Part 2, 10 relatively short chapters, covers general conventions (e.g., capitalization, punctuation, number writing, and word choice) widely used in scholarly publishing (220 pages). Part 3, 13 separate chapters covering over 300 pages, covers details of style particular to the sciences. The editors note that these chapters are

> not organized by sciences but by their subjects because the older boundaries of scientific disciplines are breaking down. Genetics and biochemistry, for example, were for many years quite separate in the subjects they covered. No longer—the molecular structure of genes and how they chemically produce their effects are subjects complexly intertwined with the older subjects of genetic inquiry. Therefore Part 3 starts with the

fundamental units of the universe and proceeds up the scale of dimensions, through chemical and cellular components, the cell, microorganisms, more complex organisms, and human society, to Earth and the rest of the universe. (CBE, 1994, p. 4)

Part 4 of the manual recommends format for scientific journals and books (154 pages), with a strong section of citations and references. The last part of the manual reviews procedural aspects of publishing.

In addition to looking at the differences in the emphases in the actual makeup of the manual, contrasting the differences in the makeup of contributors and consultants to the CBE Style Manual Committee (listed in the columns in Table 12.1) is another

TABLE 12.1.

Changes in contributors and consultants to the 5th and 6th editions of the CBE manual reflect the expanding discourse communities concerned with the language of science.

	5th Edition, 1983	6th Edition, 1994
Length	324 pages	825 pages
Contributors and consultants	42	101 (plus anonymous reviews)
Contributors from scientific journals	Editor, *Systematic Botany Monographs* Editor, *Journal of Histochemistry and Cytochemistry* Editor, *Handbook of Physiology* Editor, *Annals for Internal Medicine* Editor, *Journal of Applied Physiology* Editor, *Journal of Veterinary Research* Editor, *Journal of Heredity* Editor, *Flora Nova-Galiciana* Head, *Journals of the American Chemical Society* Editor (former), *Madrono* Editor, *Brittonia* Editor, *Anatomical Record* Publication Head, Center for Disease Control	No editorial roles named *Nature Genetics* *Science* *American Mineralogist*
Contributors from non-academic scientific organizations	National Bureau of Standards Federation of American Societies for Experimental Biology U. S. Department of Agriculture Association of American Publishers American Type Culture Collection U. S. Soil Conservation Service New York Botanical Garden BioSciences Information Service	American Chemical Society American Type Culture Collection American Mathematical Association American Physical Society American Institute of Physics American Society of Agronomy American Society for Microbiology Chemical Abstracts Service National Academy of Sciences National Cancer Institute National Institute of Health National Library of Medicine National Geographic Society National Museum of Natural History University of Chicago Press U. S. Department of Agriculture U. S. Geological Survey Australian Government Publishing Service CERN (Geneva) Commission on Plant Gene Nomenclature European Association of Science Editors International Astronomical Union International Rice Institute Institut Municipal d'Investigacio Medica Radiation Effects Research Foundation World Bank World Health Organization

way of gauging the influences of change from the 5th to the 6th edition. For example, a dozen editors of mostly U. S.-based journals are named as contributors to the 5th edition. In the 6th edition, however, no editors are named. For the 6th edition, the CBE Style Manual Committee turned away from the editorial viewpoint of discipline-specific journals to the perspectives of international organizations, from astronomers to world bankers, from Australia, Denmark, France, Japan, the Netherlands, the People's Republic of China, Spain, Switzerland, and the UK. The members of these organizations, en masse, both publish broadly and are interested in the advancement of science in both its broadest and narrowest concerns. These contributors may be seen, in part at least, not only as reflecting changes in the community of readers, but also as reflecting the broadening beat to which science marches—how science is being done differently—as the boundaries between scientific fields decompose.

In short, the changes from the 5th to the 6th edition of the CBE Manual reflect the fragmentation of science into deeply interrelated specialties. Both readers and writers can only be expected to be truly expert in a very small slice of what is becoming an increasingly interdisciplinary pie. On the one hand, then, scientists must succumb to the dynamics of the specialized language required by their expertise, while at the same time they are increasingly pressed to maintain their ability to translate and contextualize their jargon and findings to an expanding audience of interested parties. To become part of the discourse community, therefore, scientists must not only be able to write clearly to professional brethren, but also to concisely communicate the import of their findings to non-native speakers of their professional tongue.

LINGUISTIC MOORINGS IN SCIENTIFIC PUBLICATIONS: SUGGESTIONS TO AUTHORS

Although the CBE's manual is a standard reference, far from all journals in the biological sciences directly subscribe to its guidelines. Instead, all journals have their own instructions to authors, which generally cover matters of style, format, and form. Though some refer authors to the CBE for suggestions on style, many do not.

Therefore another and perhaps more important way to look at the linguistic pressure weighing on scientists is to look at variation and sameness in instructions to authors and how these specifications might cause a scientist to ponder decisions about how to construct a text.[1] In this section, I briefly describe the suggestions to authors (also called Information for Contributors and similar titles) of three journals, *Nature*, *Ecology*, and *Cell*, primarily for the sake of comparison. I chose to review these three particular journals for a number of reasons. First, these journals, in theory, focus on a narrowing range of concerns in biology, from an interest in the broadest possible spectrum of things biological in *Nature*, to a much more narrow interest in experimental investigations in biology at the cellular level and below in *Ecology* and *Cell*. Second, an author of a groundbreaking piece of research in, let's say, the biochemistry of neurons could, again in theory, consider publishing the work in any one of these journals.

In practice, this is highly unlikely to happen; if a work was shattering enough to warrant consideration for publication in *Nature*, the authors would be unlikely to choose to publish elsewhere because of the prestige that journal carries. Third, since the journals I have chosen in theory reflect a narrowing spectrum of specificity and, potentially, audience, moving from one of these journals to another could easily warrant a substantial reorientation of the presentation of the research. Some researchers might choose to send the same manuscript to another journal, rather than to revise. As an aside, I must note that in my discussion I am keeping my focus on directives about the construction of text, rather than figures, tables, and other visual representations. The issue of how data and other evidence can be presented in scientific publications is a significant one, one which warrants a separate and thorough discussion far beyond the scope of this chapter.

Nature

Nature, a weekly international publication covering all the sciences and intended for an interdisciplinary readership, is widely acknowledged as one of the most prestigious scientific publications in the world. Actually, *Nature* now has a family of international journals. In addition to the original weekly, founded in 1869, there are now four additional monthlies: *Nature Genetics* (1992), *Nature Structural Biology* (1994), *Nature Medicine* (1995) and *Nature Biotechnology* (1996). Authors of manuscripts rejected by *Nature* may be advised to try submitting their work to one of its discipline-specific offspring.

Nature's editors are extremely particular about what they publish. As mentioned earlier, the annual acceptance rate for manuscripts is only 8.5 percent. Peter Wrobel, *Nature*'s editorial manager, elaborates, "It is extremely rare (about 1 percent of accepted manuscripts) for a manuscript to be accepted without revision—that is, not be revised following referees' comments. On the other hand, not within living memory has a manuscript been accepted and published without editorial changes required by the *Nature* editors" (personal communication, November 11, 1996). The first instructions that its editors give to potential contributors is to write very concisely, as space is extremely limited and competition severe, and to write very plainly as English is not the first language of many of the journal's readers. Instructions go on to say that research articles are accepted by *Nature* editors only if they report the most novel conclusions of broad general scientific interest and only if they represent a substantial advance in scientific understanding of an important problem. These articles can be a maximum of five of *Nature*'s printed pages (6,500 words if no figures are included) and can have a maximum of 50 references. The editors require the first few paragraphs to function as both introduction and summary, and to clearly state the paper's main conclusions. Full research articles should include informative headings that reinforce the paper's logic and, therefore, should not simply include the Introduction-Methods-Results-And-Discussion (IMRAD) labels. Nonetheless, by following the scientific method, these articles often end up following an IMRAD format, although with more informative labels. In an increasingly popular trend, *Nature* requires that only brief details of materials and methods be given in the main body of the text. Instead, authors must put limited details of materials and methods at the end of the text in a section that cannot exceed 800 words. Authors needing additional space to record information needed to make their results reproducible are told to deposit the information on *Nature*'s website.

However, *Nature* often carries only one full research article in each issue. The remainder of the publication is filled with entries in the four additional categories of work that *Nature* publishes, each with its own content parameters and format specifications: review articles, progress articles, letters, and scientific correspondence. Getting one's work published in any of these categories in *Nature* is a substantial achievement. Review articles on recent developments in a vibrant area of scientific research, often longer than research reports with double the number of references, must be written for a broad readership regardless of the topic. Progress articles, a maximum of four pages, are published within a few weeks of being accepted by *Nature* editors and review rapidly developing fields in a way that allows nonspecialist readers to understand the critical nature of the progress. Letters are very short reports of original research focused on outstanding findings, which again must be digestible by nonspecialists. The first paragraph is given a maximum of 180 words, and it must summarize the background, rationale, conclusions, and implications of the work. It has a 2½ page, 30 reference maximum. Finally, *Nature* publishes scientific correspondence in what they call an informal format—extremely short reports of novel findings or preliminary results of exceptional yet general interest, or comments on other material published in *Nature*. Even these pieces of correspondence are usually peer-reviewed.

Clearly, in this journal, the IMRAD format will help researchers publish their work in only one of several possible formats, all equally viable, all weighty on a publications list. Instead, what can make or break the success of a scientist in getting into *Nature* is being able to articulate why and how findings are novel and are of both broad and deep significance. Without question, brevity, clarity, and grace will count in these situations; certainly, active voice takes less ink than passive. But knowing how to build a context for a nonspecialist will count for as much if not more.

With this said, a standard format becomes hard to prescribe. Consider for example, these two opening sections to research articles in *Nature*.

> Receptors for the Fc portion of immunoglobulins play a number of important roles in the immune system, including effector functions such as antibody-dependent cytotoxicity and phagocytosis, and transport of immunoglobulin across layers of tightly joined cells. One class of Fc receptors (FcRn) mediates transfer of maternal IgG to the newborn, providing humoral immunity to the neonate before the development of a fully functional immune system. Before weaning, suckling mice and rats express a heterodimeric FcRn in the proximal small intestine which binds maternal IgG at the pH of milk in the gut (pH 6.0–6.5). (Burmeister, Gastinel, Simister, Blum & Bjorkman, 1994, p. 336)

> Much current work is directed towards identifying molecules involved in embryonic induction and to describing their developmental effects. It is thought that some kinds of inducer molecules, acting as morphogens, are synthesized and emitted by cells in one region of an embryo and become diluted in concentration as they spread to cells distant from this point. A favoured hypothesis is that the type of gene activation and cell differentiation undergone by responding cells is related to the concentration of inducer that they receive. (Gurdon, Harger, Mitchell, & Lemaire, 1994, p. 487)

Clearly the authors of these two pieces have noticeably different styles. The first opening has tremendously long sentences and includes an enormous amount of specialized terminology. The combination of jargon and syntax throws readers directly into the meat of content, reflecting an expectation that the reader will be familiar enough with the topic to be able to comprehend. There is little orientation for the nonspecialist. The second opening, although quite technical and entirely passive, has fewer specialized terms and provides a bit more digestible context for the reader. If one could take beginnings as a reasonable sample of a text, one could reasonably predict that a broader spectrum of readers could grasp the basic gist of the second text.

A letter to *Nature* in the section definitively aimed at nonspecialist readers shows yet another equally acceptable, equally "scientific" style:

> A much debated question is whether sex differences exist in the functional organization of the brain for language. A long held hypothesis posits that language functions are more likely to be highly lateralized in males and to be represented in both cerebral hemispheres in females, but attempts to demonstrate this have been inconclusive. Here we use echo-planar functional magnetic resonance imaging to study 38 right-handed subjects (19 males and 19 females) during orthographic (letter recognition), phonological (rhyme) and semantic category tasks. During phonological tasks, brain activation in males is lateralized to the left inferior frontal gyrus regions; in females the pattern of activation is very different, engaging more diffuse neural systems that involve both the left and right interior frontal gyrus. Our data provide clear evidence for a sex difference in the functional organization of the brain for language and indicate that these variations exist at the level of phonological processing. (Shaywitz et al., 1995, p. 607)

(Note, interestingly, that all three of the passages above yield the Flesch reading grade level of 17.) *Nature* provides a good sampling of the kinds of writing tasks a researcher may face. Clearly, because of the breadth of its audience, *Nature* includes a wide range of scientific writing, perhaps more so than other publications. But, as the apex of scientific publications, its categories of acceptable prose provide more the standard than the exception. Other publications, from *Ecology*, to *Cell*, *Plant Physiology*, and *Journal of Biological Chemistry*, reflect both the mass of scientific information there is to publish, and the ever increasing population of readers interested in understanding the implications of the scientific work.

Ecology

Ecology, one of the most preeminent journals in the field of ecology and environmental science, publishes research and discussion papers developing or testing new ecological concepts or theories. Readers of *Ecology* are interested in a wide spectrum of problems, including but not limited to physiological responses of individual organisms to their abiotic and biotic environments, ecological genetics and evolution, the structure and dynamics of populations, landscape ecology, and the processing of energy and matter in ecosystems. Although not quite as severe as *Nature*, *Ecology* still only accepts about one third of the manuscripts it is sent to review.

In their Instructions to Authors, *Ecology* editors do specify some particular aspects of form and style they would prefer authors to follow, such as using the active voice and first person whenever possible. They also suggest that authors refer to the CBE manual for further guidance on style and usage. In contrast to *Nature*, the *Ecology* editors require authors to follow the IMRAD format in reports of research unless the research demands a different organization. They also specifically state that the methods section provide a complete description of the experimental design, sampling procedures, and statistical procedures for all reports of field studies, simulations, or experiments. Authors must also provide complete citations for any computer software mentioned in their texts.

However, other parameters laid out by the editors may have as much if not more impact on whether or not a manuscript is accepted for publication. Specifically, for example, the editors ask that treatment of the research topic should lead to generalizations that are potentially applicable to other species, populations, communities, or ecosystems. Robert Peet, the Editor-in-Chief of *Ecology*, states that when a report is neither of broad interest nor a critical advance in a subfield, the editors consciously try to refer the author to another publication that may be more appropriate for his or her work. “In some cases, where the science is good but not of sufficiently general interest, an international specialty journal is suggested, or where the work is more descriptive a regional journal is suggested. In other cases where we have doubts as to the quality of the science, we do not recommend publication at all, or we recommend submission to a journal that we perceive as having less stringent standards” (personal communication, August 20, 1997).

Hence, to maximize their chance of publishing a full research report in *Ecology*, researchers must be sure to state clearly the broad implications of their work and must provide evidence to explicitly back that claim in addition to adhering to the IMRAD format—if the design of their research warrants it. At the same time, *Ecology* has also recently begun a section for short papers, less tightly adhering to the IMRAD format, as well as a section for comments on papers that have been published, so researchers can nonetheless find a forum for their work and their views in *Ecology* even if it does not meet the strict criteria of being broadly generalizable.

Cell

Cell, a highly respected journal that focuses on reporting novel results in any area of experimental biology, is worthy of note for a number of reasons. First, *Cell* is known for its strict adherence to the use of active voice, personal pronouns, informative headings, and short, declarative sentences. It was one of the first publications to advocate the break from the traditional passive, pseudo-objective style often mistakenly taken as "scientific." Consider, for example, the declarative, straightforward style of this lead to the introduction from a 1994 article in *Cell:*

> A feature of all cells is the presence of bilayer membranes of a complex lipid composition. However, the role of lipid composition in biological membrane function is not well understood. Best known is the role of specific lipid-derived molecules, such as inositol phosphates and diacylglycerol, as messengers in transmembrane signal transduction. Recently, it has become increasingly clear that the regulation of lipid composition and metabolism is important to the maintenance of membrane traffic and organelle integrity. Possibly related to these issues is the potential of lipid composition to affect membrane curvature, fluidity, and fusion. (Pavlidis, Ramaswami, & Tanouye, 1994, p.23)

Second, in relation to other publications, the Information to Contributors section indicates that the turnaround time for publication in *Cell* is significantly more rapid than in many other publications, where an article can be "in press" for up to a year. The editors say that they try to let authors have reviewers' comments in less than a month after receipt of a manuscript, that revised manuscripts will be published only if resubmitted within two months of the authors' receiving reviewers' comments, and that manuscripts will be published within three months of final acceptance. This is a stark contrast to, for example, *Ecology*, where a decision on acceptance is generally rendered within three to four months of receipt of a manuscript, but can be longer. Authors are given up to four months to resubmit revised manuscripts. An article usually appears in *Ecology* about four months after the final manuscript has been approved.

Third, as a result of their rapid turnaround of full reports, the editors are absolute in some of their directives. In the information to prospective authors, the editors state

> The total character count of an article must be under 55,000. There should be no more than 7 figures ... Each figure should consist of a single montage that can be reduced to one column width. Papers that do not fit this format will be returned to the authors. ... Papers that do not conform to the general criteria for publication will be returned immediately, without detailed review. (*Cell*, 1997)

In other words, in addition to making the novel significance of his or her work clear and explicit to readers outside his or her immediate range of expertise (as most of the editors of *Cell* are likely to be for any one particular manuscript) authors wishing to publish in *Cell* must also construct the physical text according to prescribed guidlines, or else, the article will not even be sent out to members of the editorial board for review.

Last, in a significant departure from many other publications focusing on the presentation of experimental results, *Cell* also allows authors to combine the presentation of their results with discussion. Having the opportunity to interpret results in the context of presenting them is not usual and potentially allows for a different kind of argument to be built than would be possible if the presentation and discussion of data were strictly separated. In addition to full reports, *Cell* publishes minireviews and letters; authors contemplating a minireview must submit and have approved a proposal for the review before they can even submit a full manuscript.

TEACHING DISCIPLINARY WRITING SKILLS

This review of suggestions to authors is meant to underscore that in the modern world of publishing science, particularly in the biological sciences, knowing the nuts and bolts of what information should go in which section of a formal scientific report barely scratches the basics of competency in written communication skills. Without question, knowing and understanding the IMRAD format is important. Knowing how to compose stylistically streamlined sentences and fearlessly practicing the construction of active and declarative prose perhaps can help. However, an increasingly necessary skill is knowing how to argue for and situate the significance of the findings for different audiences in different forums. The full research report is only one of a number of different forms through which scientific findings can be publicized, and it seems that the more widely read and prestigious the publication, the more important it is that scientific results be articulated in a contextualized and digestible form.

Beyond the IMRAD report, then, scientists in training need to become practiced in making decisions about what, where, when, and why to publish their findings. Should one sit on a particular set of revealing data, for example, or send it in for a rapid communication? Should one construct one's prose to aim to publish in a prestigious journal with a wide readership, a journal that insists on novel and widely appealing results, or should one write up the work in a more conservative fashion for a less competitive publication, aiming just to get into print?

Becoming aware of and practiced in making these kinds of rhetorical decisions is not only important for researchers who want to publish within the academic community but also for those who need to communicate about science in a broad spectrum of applied settings. Although many newly emerged PhD biologists strive to emulate their teachers and academic mentors, the academic job market is only so big. At times more than a few recent graduates migrate into industry by default—or by design—to positions where clear, concise writing about the research is essential for their success, since many readers will be nonspecialists (often managers) who are interested in the applied product of the science rather than the theory, model, or process that led to it.

What the modern demands of scientific writing suggest for teachers is twofold.

First, the demands suggest that scientific writing should be taught as an integral part of teaching science; a writer can only come to understand more profoundly both a problem and its potential solutions by learning to articulate the problem and argue for solutions to readers with different backgrounds and orientations. During the process of articulating problems and solutions, the writer comes to understand the audience more intimately, and in so doing he or she grows deeper roots into the discourse community. In the biological sciences at least, it seems that the expectations of the discourse community are evolving to value scientists' abilities to show the interconnectedness of what one does with what others are doing, to show novelty and sameness at the same time.

Second, the demands of modern scientific writing suggest that to teach science writing well, or at least in the ideal way, a scientist and a rhetorician (or at least a non-deconstructionist composition instructor) need to work together. The scientist can learn some of the skills of the rhetorician—that ability to find a means of persuasion in all things. The rhetorician in turn can learn more specifics about that which must be considered in communicating the particularities of a slice of science, and in so doing can learn new ways of provoking change by infusing the notions of purpose and audience into what is often otherwise taught and learned by rote.

These suggestions to teachers are sound, I believe, but extremely difficult to achieve. A variety of pressures work against the incentives to teach rhetorical and critical thinking skills as part of writing instruction in the sciences. One such pressure is made of the mixture of time and tenure politics: Active scientists on tenure tracks are generally not given points toward tenure for teaching courses that stress writing, or for collaborating with those outside their scientific discipline. Another pressure faced by curriculum developers in scientific disciplines is how to squeeze in all the content courses students need in order to graduate in four years with the fundamentals of "Biology" or "Chemistry" under their belts. Curriculum developers faced with assigning a teacher to develop a content-based course or a writing-intensive one that uses content as its fodder will probably choose the content course, as it is a route that is easier to understand and justify.

Teachers, then, who want to develop courses in collaboration with teaching scientists will need to work slowly, providing all involved with small pieces of evidence that teaching writing in content courses will help, not detract from, the students' learning of content and can help them understand content more deeply. As importantly, teachers will also have to provide convincing evidence that teaching a writing-intensive course will not add significantly to the workload of the scientist. Since many scientists do not rank writing as one of their favorite tasks, this prerequisite for developing a teaching relationship between writing teachers and science researchers is at the same time the most important and the most difficult for building functional collaborations.

ACKNOWLEDGMENTS

I would like to thank Drynda Johnson of the University of Pittsburgh Science Library for helping me with investigations leading to this study, as well as Dr. John Hempel and Dr. Laurel Roberts of the University of Pittsburgh Department of Biological Sciences for their reviews of earlier versions of this manuscript.

NOTE

[1] For a sampling of hundreds of suggestions to authors from many scientific publications related to biological and medical sciences, interested readers can visit web sites such as the index to Instructions to Authors in Health Sciences put out by the Medical College of Ohio [*http://www.mco.edu/lib/instr/libinsta.html*] or Pedro's Biomolecular Research Tool index of bio/chemical journals and newsletters [*http://www.public.iastate.edu/~pedro/rt_journals.html*].

REFERENCES

Angier, N. (1995, January 24). Sex and the fruit fly: Price of promiscuity is premature death. *New York Times*, pp. B7–B8.

Booth, V. (1994). *Communicating in science* (2nd ed.). New York: Cambridge University Press.

Burmeister, W., Gastinel, L., Simister, N., Blum, M., & Bjorkman, P. (1994). Crystal structure at 2.2°: A resolution of the MHC-related neonatal Fc receptor. *Nature, 372,* 336–343.

Cell. (1997). Information for Contributors [World Wide Web]. Available: http://www.cell.com/misc/itoa.shtml.

Chapman, T., Liddle, L., Kalb, J., Wolfner, M., & Lartride, L. (1995). Cost of mating in *Drosophila melanogaster* females is mediated by male accessory gland products, *Nature, 373,* 241–244.

Council of Biological Editors Style Manual Committee. (1994). *Scientific style and format: The CBE manual for authors, editors, and publishers* (6th ed.). New York: Cambridge University Press.

Day, R. (1992). *Scientific English: A guide for scientists and other professionals.* Phoenix, AZ: Oryx.

Day, R. (1994). *How to write and publish a scientific paper* (4th ed.). Phoenix, AZ: Oryx.

Document Supply Centre. (1996, January). *Current Serials Listing.* London: British Library.

Ebel, H., Bliefert, C., & Russey, W. (1990). *The art of scientific writing.* New York: VCH.

Gurdon, J.B., Harger, P., Mitchell, A., & Lemaire, P. (1994). Activin signalling and response to a morphogen gradient. *Nature, 371,* 487–492.

Lavoisier, A. L. (1965). *Oeuvres de Lavoisier Publiees par les soins de son excellence le ministre de l'instruction publique et des cultes* [The works of Lavoisier published under the auspices of His Excellency the Minister of Public Education and Worship]. New York: Johnson Reprint Corporation. Reprinted from *Oeuvres de Lavoisier Publiees par les soins de son excellence le ministre de l'instruction publique et des cultes,* by A. L. Lavoisier, 1864–1893, Paris: Imprimerie Imperiale.

McMillan, V. (1997). *Writing papers in the biological sciences* (2nd ed.). Boston: Bedford.

Pavlidis, P., Ramaswami, M., & Tanouye, M. (1994). The Drosophila easily shocked gene: A mutation in a phospholipid synthetic pathway causes seizure, neuronal failure, and paralysis. *Cell, 79,* 23–33.

Pechenik, J. (1993). *A short guide to writing about biology* (2nd ed.). New York: Harper Collins.

Shaywitz, B., Shaywitz, S., Pugh, K., Constable, R., Skudlarski, P., Fulbright, R., Bronen, R., Fletcher, J., Shankweiler, D., Katz, L., & Gore, J. (1995). Sex differences in the functional organization of the brain for language. *Nature, 373,* 607–609.

13

Textual and Contextual Models of Science Writing

Ramón Plo-Alastrué
Carmen Pérez-Llantada
University of Zaragoza

The aim of this chapter is to describe several criteria for adopting certain types of texts as models of science writing in an English for Science (ES) course for Spanish university students. In selecting models appropriate to the special needs of our students, we have focused on one particular characteristic: the students' reasonably good knowledge of the scientific content together with a still poor command of the specialized language in English. Finally, and apart from the traditional textual examples, we propose a series of contextual models, which may help familiarize prospective scientists with some common schemes and practices of their discourse community. Both textual and contextual models together have an important role in developing the students' communicative competence, implying a knowledge of grammar, vocabulary, and semantics, as well as the patterns of sociolinguistic behavior of the speech community.

As it is easily traceable across contemporary theories, studies, and trends, English for Specific Purposes (ESP) has become one of the areas of applied linguistics which is receiving increasing attention at present. Among ESP's different branches, English for Science and Technology (EST) has undergone the most rapid development. On many occasions, ESP and EST have been taken almost as synonyms because most of the early investigations on the subject took EST—an apparently "extreme" register—as their main object of study. Several authors later contributed to establishing its special importance by setting up a series of classifications where they distinguished EST from "all other courses" (Strevens, 1977, p. 90). Since Barber's analysis (1962)—usually taken as the starting point of the modern concept of ESP—there have been many different attempts to focus, define, and teach what EST is in terms of Register or Frequency Analyses (Biber, 1988), grammatical-rhetorical analyses (Jordan, 1984) or, for example, the so-called "learning-centered" approach (Hutchinson & Waters, 1987).

Although science and technology were initially taken as a unitary field, it is possible to see at present an increasing interest in underlining the differences between both areas and, from that double perspective, to start a more precise study of the different disciplines included under this general label. Strevens (1980), for example, provides a distinction based on their object of study and defines science as "concerned with understanding, describing, and explaining the nature of the universe (including, of course, Man)" whereas technology "is concerned with how to design, operate, and control machines, devices and instruments" (p. 128). This specialization of tasks demands the existence of a third element—"technical services"—in charge of constructing and maintaining "the devices invented by technology according to the principles established in science" (p. 128). Other authors, like Biesenbach-Lucas (1994), choose the communicative function as the distinctive trait. Quoting the work of Stratton (1979), she underlines and justifies this distinction by saying, "What appears to differentiate technical writing from scientific writing ... is a pragmatic intention of the former, which is designed to elicit some overt behavioral response on the part of the reader ... and which is absent from scientific writing" (p. 33).

We think, however, that this absence of pragmatic intention is not so clear, even though it has been commonly stressed in order to convey that scientific writing is purely referential. The problem is that scientific prose traditionally has been characterized by its objectivity in the presentation of ideas and its reliability in the exposition of facts, statements, and judgments. It is meaningful that this generalized view—although more and more questioned—tends to be maintained mostly by scientists themselves, rather than by those who study scientists and their discourse (Valle, 1995).

In this sense, Harré (1990) considers that the scientific community must be seen as a moral order, a kind of élite "whose internal structure is based upon a network of trust and faith" (p. 82). Some of the linguistic features displayed in a science discourse in order to achieve that faith are, according to Gilbert and Mulkey (1984), "the use of an impersonal style, with references to the author's actions and judgments kept to the minimum" (p. 56). The combination of these elements creates the impression of unmediated contact with a physical world which is not being interpreted but rather speaking directly to us in an almost religious experience. Apart from these linguistic features, there are also rhetorical or narrative conventions which greatly contribute to this final impression. One of them is the presentation of scientific research in such a way that it wins the confidence of the reader or listener. Everything described by published papers works out as expected—even though, sometimes, this account bears little or no resemblance to what actually happens in real life. The result is what Harré calls a "smiling face" presentation of scientific research. In other words, what we get is not the real story but a way of telling it that turns it into a "piece of fiction" (pp. 86–87).

From this perspective, scientific prose appears as the result of a complex process of transformation where the events are selected, their order is altered, and the whole operation is described through certain linguistic features, narrative conventions, and strict rhetorical norms. Paradoxically, all these elements render this same process invisible—the apparent "absence of pragmatic intention" mentioned above—thus creating the impression of unquestionable objectivity and authority.

Another complementary view is that whereas science concentrates on strict patterns for the presentation of new data, procedures, and methods of research to a specialized audience, technology is more interested in informing both specialists and nonspecialists about the latest advances in the field. It is, in short, the functionality—understood as the degree of scientific and technical specialization or knowledge of the target audience—that distinguishes the discourse of science from the discourse of technology. This distinction introduces a new and important idea: There is a variety of social and linguistic issues that must also be taken into account. Thus, apart from giving the impression of being objective and reliable, scientific discourse must also be both effective in the purpose of the communication and appropriate for the audience addressed.

In any case, and apart from these aspects, a closer view of the discourse for science would necessarily remind us of some other discoursal concepts such as patterns (Jordan, 1984; Hoey, 1983), rhetorical functions and techniques (Trimble, 1985), and semantic relations between propositions (Crombie, 1985). Our proposal, however, is that both linguistic and rhetorical parameters are shaped according to the communicative nature of language and are, thus, directly dependent on its pragmatic dimension.

As for the most recurrent written material handled by both academics and professionals in the fields of technical engineering, a needs analysis carried out in our university showed that research articles, reports, instructions, technical advertisements, and formal letters were the most widely used genre-types in this particular university community. On the other hand, the needs of scientists can be described as basically academic and therefore the most frequently used genre-types are those related to the area of education and research. However, as we will later see, if we consider some further typologies of scientific genres, it will be necessary to include in our pedagogical approach some references to the pragmatic aspects underpinning—as also happened with technical writing—science writing.

There are, in fact, some general characteristics of the scientific prose, which are based on the fact that scientists need to write as a basic need. According to Day (1979):

> The research scientist, perhaps uniquely among the trades and professions, must provide a written document showing what he or she did, why it was done, how it was done, and

> what was learned from it. The key word is *reproducibility*. That is what makes science and scientific writing unique. (p. ix)

This institutional pressure on the scientist to make his/her results public (and, if possible, sooner than the other colleagues working on the same subject) has been summarized under the expression "publish or perish." In any case, this general need to publish is especially troublesome for nonnative students of science who—leaving aside the complexities of content—have the extra difficulty of learning both a new specialized language and also the discoursal practices of this particular scientific community.

Our students' present profile can be described using the following three elements. They are university students with a reasonably good knowledge of scientific content, an elementary knowledge of the frame of mind of the scientific community, and a poor command of scientific English (especially the productive skills).

On many occasions, this lack of fluency in the specialized language has been chosen as the only, or at least the main, problem to tackle. For example, Zimmerman, in his book *English for Science* (1989), describes this common attitude:

> The non-native speaker may be familiar with the science concepts that are covered but not know how to express them in English. The focus of the book, however, is always on language rather than science, and there is no attempt to cover any scientific topic thoroughly. (p. xix)

Based on this conception, we can see a whole series of EST courses where vocabulary and semantics are extensively revised together with the appropriate grammar and rhetorical elements. However, in these same texts, the sociolinguistic or pragmatic dimension is systematically neglected and the scientific content considered—from a typically nonscientist point of view—a kind of "secondary difficulty," instead of a source of motivation or help.

We think, however, that these three elements should not be seen as independent, but as interrelated aspects. Thus, the higher the degree of scientific knowledge required to interest students, the more constricted and linguistically patterned a given text will be and the higher its linguistic level of difficulty (e.g., abundance of nominalizations, etc.).

Paradoxically, this increase in the complexity of the texts, and tendency toward a global (semantic, linguistic, rhetorical, and scientific) "specialization" does not necessarily mean that our students find these texts more difficult. In many cases, it is precisely this specialized vocabulary and normative linguistic patterning that can be of great help in extracting the meaning of a text or making the model easier to imitate. From this perspective, for example, science students usually find that the highly constrained "Methods and Materials" section of a research article is much easier to write than the final "Discussion," where the specialized vocabulary is not so defined and there is room for expressing their personal interpretation of the facts and comparing them to the literature on the subject. It is not then a simple question of choosing between content (and motivation) and form (and level of difficulty) but of finding the right balance between these aspects and choosing the level that meets the specific needs of our students.

Logically, the first objective of most EST courses is to cater to the primary need of all students: reading comprehension of the most important types of science literature. Since English has become the language of science, this task is seen as unavoidable and students' motivation is high. Because our experience tells us that most scientists have a good command of this skill, we decided to start our science writing course with these reading comprehension skills and analyze textual models of several scientific genres.

The analysis of appropriate models has been one of the methods traditionally used in writing courses. In his classical book *Writing Scientific Papers and Reports*, W. P. Jones (1946) recommended the use of models to learn basic skills such as how to organize and present facts clearly and systematically, how to use specific facts to support a general conclusion, or how to distinguish between facts and opinions. By handling in class the different typologies of science writing, students become aware of the communicative nature of the language they are learning, and they learn to schematize the minute stratification—both at a textual and at a contextual (discourse) level—of this particular discipline.

There are well-known recent proposals discussing the key features of science writing which can be of great help, such as the ones given by Day (1979), Huckin and Olsen (1983), or Swales (1990). Among the different genres, the list of chosen models was restricted to our students' needs and profile, which can be described as basically academic. For that reason, we have focused on the so-called "research-process genres" (Swales, 1990, p. 177) characteristic of science writing: Thus, apart from the research article, we can include in this list the abstract, theses and dissertations, and books and monographs.

TEXTUAL MODELS

The list of models chosen from the research-process genres is not exclusive but covers the basic academic writing needs of our students at a textual level. In general, the first and main advantage of these authentic materials is that they are relevant, and therefore motivate students. However, their use also has clear disadvantages, such as the lack of a clear pedagogic gradation or of a primarily linguistic orientation. In this process of grading materials, we have adopted the three criteria put forward by MacMillan (in McDonough, 1984, p. 69)—science content, language content (elementary, intermediate, and advanced) and use (spoken informal and formal, written, etc.), including in this last criterion the textual organization or formal structure of the text in question.

Another factor taken into consideration was the communicative purpose; that is, the possibility of drawing on the knowledge of one type of text in order to analyze the following type. From this perspective, we established the following gradation:

1. Textbooks
    1.1. Basic
    1.2. Monographs
2. Theses and dissertations
3. Abstracts
    3.1. Abstracts in journals and proceedings
    3.2. Extended abstracts/summaries
4. Research article
    4.1. State-of-the-art (or review article)
    4.2. Indexed article
    4.3. Popular science

The first models introduced in the course are textbooks. Although textbooks traditionally have been left aside as a "minor" genre, their important role in the scientist's formation is admitted by more and more professionals. In a recent article, Swales (1994) describes textbooks as "deserving members of the academic genre-system" (p. 16), arguing that they may "consolidate and apply recent scholarship, incorporate new research findings, and generate interesting new topics worth further study" (p. 3). Their main role in our course is as a kind of introductory genre to the specialized language of science through texts which deal with scientific concepts from a more didactic perspective.

Theses and dissertations, the second class of texts, have been mostly introduced as models of scientific style as well as of textual organization. This limited role is likely to change in the near future since science students in Spain have started writing and defending these texts in English. It is interesting to note, in this sense, that these types of texts have been perceived as the primary texts needed by international graduate students at British and American universities.

After this gradual introduction to several characteristics of science discourse provided by the first two models, the third part of the course is devoted to the research article and the abstract. Knorr-Cetina (as cited in Swales, 1990, p. 95) describes the research article as "the standard product of the knowledge-manufacturing industries." The reason for this model's importance is not only its predominance but also its specialization: The scientific article is the common vehicle chosen for validating the latest experiments, advances, and results in scientific research. This high degree of specialization can also be seen in its scope—most articles are based on a single investigation, thus giving an idea of completeness—and in its audience, which consists of specialists in the field. Partly as a result of their search for effective specialized communication, research articles have adopted strict norms and formal conventions, such as the well-known IMRAD structure, which students learn and understand not as something imposed by rhetoric but as the gradual result of the influence of its function, audience, and context.

As for the three criteria applied for creating a gradation, the models of textbooks chosen for the course show a parallel increase in the linguistic level of difficulty and the complexity of scientific content without great changes in their formal structure. Abstracts are included as an autonomous genre whose audience, context, and function—informative, descriptive, or a mixture of both (O'Connor, 1991, p. 70)—radically affect their linguistic patternings. The research article, in turn, could be described as having a rigid formal structure and a high degree of specialization both on the linguistic level and in its scientific content. At the other end of the scale, the formal structure of a popular-science article is less rigid: Its scientific level of specialization is lower but its linguistic level of difficulty becomes much higher for nonnative students.

In general, students prefer texts with rigid formal structures since the more constricted —or more structurally and linguistically patterned—the text is, the higher the degree of scientific knowledge required and vice versa.

Paradoxically, these students find no extra difficulty in texts with a high level of scientific specialization, whereas their rigid formal structure makes them easier to follow as writing models for the course. In this sense, the linguistic simplification inherent, for example, in scientific books for children makes them attractive but difficult to follow, and therefore inadequate materials for our course.

CONTEXTUAL MODELS

It can be seen that the previous list of textual models—which is common to most science writing courses—reflects the traditional idea of science as something purely objective, crystallized, and unchanging. Thus, despite the fact that in the 1930s, Fleck (1935) maintained that scientific documents were not objective descriptions of uncontested facts, there are still many who believe that this process of negotiation of meaning should be excluded from course content, considering instead only the final product: scientific writing as a transparent attempt to reflect objective truth.

However, it is important to underline the role of the influences that shape the text and to make students aware not only of the final result but also of the process—that is, of scientific procedure. In this way, they can compare their own discourse practices with those of the target community and learn both how scientists frame their research and the complete process of creating a scientific text.

From this perspective, we consider the context as a system of other textual structures closely related to the text and forming what Barthes (1977) describes as "a tissue of quotations drawn from the innumerable centres of culture" (p. 146). They are all influences that shape the text. Some of them very actively give origin to citations, repetitions, comments, or transformations in the final product while others are latent or even nonidentifiable influences. This whole set of textual antecedents is what we call, using Genette's terminology (1982, p. 11), "hypotextes," or underlining its usually previous existence, pretexts. From this perspective, we arrive at a global, more fluid concept of text immersed in an intertextual dimension which includes "everything, be it explicit or latent, that relates that text to others" (Genette, p. 7).

If one of the main characteristics of technical writing is its pragmatic intention (eliciting a response from the reader), science writing can be analyzed from this perspective as a writer's reaction to all these pretexts. Among these pretexts, we can mention three types of complementary models for our course:

1. The first is a series of published texts which are never overtly quoted by scientists but which exert an important influence on both the content and the style of their works. For example: citation indexes, the style guides and editorial policies of different journals, all kinds of reference books, etc.

   Citation analysis has been increasingly used in Spain as a method of evaluating and ranking professionals and institutions. If a paper or author is frequently cited in certain specialized journals (those with a high "impact factor"), this means that the article was well received, was used by some other scientists and was influential in its field. It is only natural, then, that many scientific articles are conceived and structured with these particular journals in mind. From this perspective, the citation index can be defined as a powerful pretext which controls many different aspects—such as the approach, the content, and the form—of the final article.

   As for the importance of the style guides provided by these journals, we only have to quote the following advice given to students in a manual on science writing:

   > Before you start to draft a paper, read the most recent set of instructions or guidelines for authors prepared by the journal, institute or agency for which you are writing. Those instructions or guidelines take precedence over any other advice, including the advice given in this book. (O'Connor, 1996, p. x)

2. We also find it interesting to include in our course some examples of the "bureaucratic" side of scientific discourse (e.g. grant proposals, application forms, reprint requests, etc.).

   The reason for including these texts is to make students familiar not only with the pure work of investigation but with a series of chores which also characterize the work of a scientist. Many students are now able to see the importance of this bureaucratic aspect of science since the functional target of all this literature lately has been foregrounded in Europe by the popular Erasmus and Socrates[1] program interchanges.
3. And, finally, we added a series of unpublished materials (e.g. personal letters, e-mail messages, calls for papers, discussion lists on the Internet, etc.) considered of secondary importance and systematically neglected in the teaching of science

writing but which can be of great help for motivating students and introducing them to the specialized language of science and to the practices of this discourse community.

In general, when faced with these contextual materials, students find them as relevant and motivating as the "classical" models of scientific texts. In addition, all these examples entail something more than a mere imitation of the proposed model: They demand a written answer (e.g., application form, reply to e-mail, etc.); that is, a real communicative activity.

Of particular interest to us in this last section of the course is the category of miscellaneous correspondence writing, as it is perhaps one of the genres most highly communicative in nature in which many of the elements introduced during the course can be seen in a practical perspective. A letter or an e-mail message is considered a mode or means by which a message is transmitted. As opposed to more formal genres such as the research article or report, for example, the main characteristics of this type of genre are its hortatory nature, its expressive rather than strictly referential manner and, above all, its interaction, as it obviously involves the reader's response to the message.

Writing for a specific purpose or intention requires organizing one's thoughts and recasting them, using the correct grammatical and organizational structures. In addition to this, written correspondence creates its own context and therefore has to be fully explicit; no immediate feedback is possible, and the writer may try to anticipate the reader's reactions and incorporate them into the text.

The final step to be considered by the writer is to adapt his/her message to the accepted style of discourse or etiquette. A letter or document must conform to the standards of social acceptability so as to meet the linguistic demands of the situation in which language is used. Differences of social situation are reflected in the choice of language according to the relationships between the writer and his/her readership. For example, in contrast to the colloquial or informal language between colleagues, we are likely to use formal language when we talk to strangers, to unknown audiences, or to people of a high status. In other words, degrees of formality will be considered the linguistic reflection of recurrent characteristics in the use of language in specific situations.

Considering the above mentioned points from a teaching perspective, one can conclude that the genre of correspondence can become an effective means of engaging students in the knowledge of sentential structure as well as in the standard formats of information transfer—namely, fixed linguistic patterns according to the communicative usefulness and the social appropriateness of the message. Only in this way will the final product be an organized piece of information with effectiveness and appropriateness of style.

The existence and increasing importance of all these contextual models has been noted by Swales (1994) in the so-called "interstitial genres" where he includes "proposals, reader's reports, editorial correspondence, etc." which in his opinion "have a major role in shaping both the process and the final public product" (p. 16). As for teaching these models, Hutchinson and Waters (as cited in Jones, 1990, p. 91) suggest the creation of a bank of materials for ESP courses, "arguing in particular for a greater variety of text types" where, in our opinion, such contextual models could be added as an important complement to the traditional genres.

It can be seen that the study of the different textual categories or genres in our writing course foregrounds the relevance of functional and interactional criteria when writing scientific texts (see also McCarthy, 1990; Sinclair, 1991; Wallace, 1992). In particular, Swales' (1990) analysis of genres represents a positive theoretical attempt to emphasize the strong relationship of grammar to context and, above all, to discourse.

The purpose of approaching our course from a genre-based perspective is to raise students' awareness of the constraints and variations of functional and situational parameters in the different genres of scientific writing. This awareness basically amounts to having a clear image of the audience and the communicative purpose of the writing—that is to say, the context of these texts. Students are first asked to organize relevant meanings cohesively and then to adapt the style of the text to its context of use since, in fact, linguistic patternings are constructed for every different genre type, shaped in response to the context of situation within which the genre is framed. Thus, we consider this awareness the key point on which all the other important writing decisions, such as paragraph creation and distribution, and elements of grammar and style, will be made.

This perspective rests primarily on the idea that students learn not only scientific content but also the vocabulary, rhetorical patterns, methods of research, discoursal practices, etc., of the target community of experts (Kuhn, 1962). There are more recent approaches which consider that this process takes place in both directions. According to these theories, since learning can be understood as an interactive process between novices and experts, it is described "not as a mental event internal to an individual but as a social achievement within a complex framework of community, goals, tools, and activities" (Jacoby & Gonzales, 1991, p. 150). If we then take into account this view of language as a societal construct, it would be worth remembering the notion of "discourse community" (Foucault 1970; revisited in Swales 1990), as any text which must ultimately conform to a given situation within a given social setting.

It is context—in the sense of purpose and situation—that determines what a text really means; as Mey describes it (1993, p. 31; emphasis added), "The meaning of an utterance is determined in large part by how it responds and how it is responded to, by its place in an interactional sequence, namely, a *context of use*." The fact that every text is thus context-bound is closely linked to the idea of the social context of language use, and such a context precisely presupposes the existence of a particular community with its implicit cultural rules. As language is patterned toward social ends, meaning is realized in the text and likewise shaped in response to the context of the situation in which it is used.

Language is then seen not only as a system of structurally related elements for the coding of meaning, but rather as a social act grounded on the principles of conventionality and appropriateness; it is, in short, what Kroll (1993) refers to as "sociolinguistic behaviour" and "adequacy of style." In this sense, we think that together with a good review of the core of the language, students need extensive written training in order to integrate the linguistic standards of their particular community.

With this purpose in mind, our course is designed to enable students to write convincingly the different forms of academic writing with great fluency. In order to do so, students are led through the process of writing with attention to logic and patterns of organization as well as to the relevance of purpose and audience, the latter considered as the two main factors that determine the linguistic choices of a given text. More particularly, students are taught to see not only how communicative components are realized in the text so as to allow for appropriate social interaction, but also how rhetorical structures and models are used to carry out a set of given communicative goals.

We thus conclude that both the description of the linguistic system together with all the social underpinnings of a discourse community should be thoroughly considered in a writing course, since the web of conventions that exists in a community demands that students know the rules for using statements in different contexts of linguistic use.

We therefore suggest as a starting point a content-based approach, which leads students through the process of writing with attention to logical and linguistic patterns of organization and development respectively, and then a pragmatic-based approach, leading to consideration of other concepts such as functional paradigms of purpose and audience. Only in this way can one raise the learner's awareness of the importance of the nature, structure, and use of language, and the influence of these in the functioning of a language; this is precisely the taxonomic principle underpinning our teaching.

While helping students to progress gradually, we attempt to make them feel confident and motivated by varied and flexible material. The use of this kind of material opens a wide range of teaching methodology, which makes available all the resources of modern language teaching techniques in an accessible and systematic form. It is our main goal to offer a revision of the essential grammar so as to provide students extensive opportunities to express their own ideas and opinions, while developing the skills to communicate effectively in English, with a general appropriateness of language and with the proper etiquette. This explains our attempt to develop in students the so-called "communicative competence" (Richards, 1985, p. 145), which entails not only the need for a correct grammatical and semantic understanding, but also the knowledge of the sociolinguistic behavior of the speech community.

NOTES

[1] Socrates is a European Community program of cooperation among universities. In a wider context it should be regarded as a context to expand "learning" as the only feasible answer to the challenge brought about by the technological changes and their enormous influence both on social knowledge and on individual potential. So as to strengthen the European influence in educational matters, this program (which also includes the so-called Erasmus program) covers a wide range of activities within the institutional contract: mobility of students, mobility of teaching staff, intensive programs, European Credit Transfer System (ECTS), and the making up of parallel syllabi. Moreover, independently of the institutional contract, projects of university cooperation in disciplines of common interest are also carried out.

REFERENCES

Barber, C. L. (1962). Some measurable characteristics of modern scientific prose. In *Contributions to English syntax and philology* (pp. 21–44). Göteborg, Sweden: Almqvist & Wiksell.

Barthes, R. (1977). The death of the author. In R. Barthes, *Image-music-text* (pp. 142–148). New York: Hill and Wang.

Biber, D. (1988). *Variations across speech and writing*. Cambridge: Cambridge University Press.

Biesenbach-Lucas, S. (1994). *A comparative genre analysis: The research article and its popularization.* (Available from UMI Dissertation Services, 300 N. Zeeb Road, Ann Arbor, MI 48106, order no. 9511259)

Crombie, W. (1985). *Process and relation in discourse and language learning*. Oxford: Oxford University Press.

Day, R. (1979). *How to write and publish a scientific paper*. Cambridge: Cambridge University Press.

Fleck, L. (1935). *Genesis and development of a scientific fact*. Chicago: University of Chicago Press.

Foucault, M. (1970). The order of discourse. In R. Young (Ed.), *Untying the text: A post-structuralist reader* (pp. 48–78). London: Routledge.

Genette, G. (1982). *Palimpsestes: La littérature au second degré*. Paris: Éditions du Seuil.

Gilbert, G. N., & Mulkey, M. (1984). *Opening Pandora's box: A sociological analysis of scientific discourse*. Cambridge: Cambridge University Press.

Harré, R. (1990). Some narrative conventions of scientific discourse. In C. Nash (Ed.), *Narrative and culture* (pp. 81–101). London: Routledge.

Hoey, M. (1983). *On the surface of discourse*. London: Allen & Unwin.

Huckin, T. N., & Olsen, L. A. (1983). *English for science and technology: A handbook for non-native speakers*. New York: McGraw-Hill.

Hutchinson, T., & Waters, A. (1987). *English for specific purposes: A learning-centered approach*. Cambridge: Cambridge University Press.

Jacoby, S., & Gonzales, P. (1991). The constitution of expert-novice in scientific discourse. *Issues in Applied Linguistics, 2,* 149–181.

Jones, W. P. (1946). *Writing scientific papers and reports*. Iowa State University: W. M. Brown.

Jones, G. M. (1990). ESP textbooks: Do they really exist? *ESP, 9* (1), 89–93.

Jordan, M. (1984). *Rhetoric of everyday English texts*. London: Allen & Unwin.

Kroll, B. (1993). *Second language writing: Research insights in the classroom*. Cambridge: Cambridge University Press.

Kuhn, T. S. (1962). *The structure of scientific revolutions*. Chicago: University of Chicago Press.

McCarthy, J. (1990). *Vocabulary*. Oxford: Oxford University Press.

McDonough, J. (1984). *ESP in perspective: A practical guide*. Glasgow: Collins.

Mey, J. L. (1993). *Pragmatics: An introduction*. Oxford: Blackwell.

O'Connor, M. (1991). *Writing successfully in science*. London: Chapman & Hall.

Richards, J. C. (1985). *The context of language teaching*. Cambridge: Cambridge University Press.

Sinclair, J. L. (1991). *Corpus, concordance, collocation*. Oxford: Oxford University Press.

Stratton, C. R. (1979). Technical writing: What it is and what it isn't. *Journal of Technical Writing and Communication, 9*, 59–65.

Strevens, P. (1977). *New orientations in the teaching of English*. Oxford: Oxford University Press.

Strevens, P. (1980). *Teaching English as an international language: From practice to principle*. Oxford: Pergamon.

Swales, J. (1990). *Genre analysis*. Cambridge: Cambridge University Press.

Swales, J. (1994). The role of the textbook in EAP writing research. *ESP, 13* (3), 3–18.

Trimble, L. (1985). *English for science and technology: A discourse approach*. Cambridge: Cambridge University Press.

Valle, E. (1995). [Review of the book *Understanding scientific prose*]. *ESP, 14* (2), 179–181.

Wallace, C. (1992). *Reading*. Oxford: Oxford University Press.

Zimmerman, F. (1989). *English for science*. Englewood Cliffs, NJ: Prentice-Hall.

14

Practice-Centered Approaches to Literacy: Applications for the Research and Teaching of Science Writing

Rosemary Horowitz
Appalachian State University

Definitions of reading and writing are changing. Until about 15 years ago in the United States, reading and writing was defined primarily in terms of cognitive and linguistic skills and in relationship to school learning. Student skills were at the center of this definition, and from this skill-centered point of view, researchers framed questions for inquiry. However, new definitions of literacy have emerged that frame reading and writing in terms of social practices rather than school learning. Instead of focusing on the cognitive skills of a single reader or writer, practice-centered approaches emphasize the roles, uses, values, attitudes, and meanings of language activities. This change reflects an understanding that literacy exists in many settings, assumes many forms, has multiple uses, has many meanings, serves many purposes, and is dynamic. It also recognizes that literacy is shaped by historical, economic, cultural, educational, and other conditions.

My goal in this chapter is to discuss the ways in which practice-centered approaches to literacy may be applied to the study and teaching of science writing. To do so, I first review concepts from various practice-centered approaches. Then, by way of illustration, I consider four examples of science writing in order to demonstrate the applicability of these perspectives. I end by discussing the implications for research and pedagogy.

WRITING AS A SOCIAL PRACTICE

Central to practice-centered approaches to literacy are five basic concepts. One concerns *sociality.* To the extent that reading and writing affects interpersonal relations, practice-centered approaches define reading and writing as a social process. For example, in their analysis of language use, Halliday and Martin (1993) ask "What is taking place?", "Who is taking part?", and "What is the role of language?" Similarly, Heath (1983), Hymes (1974, 1986), and Street (1984, 1993) look at the social aspects of language, employing the terms "uses of literacy" to denote what people do with reading and writing, "functions of literacy" to refer to the purposes served by reading and writing, and "meanings of literacy" to refer to the values, attitudes, and beliefs assigned to reading and writing. Thus, on the whole, practice-centered approaches recognize literacy as a social act and provide insights into the ways people use literacy to accomplish their goals and the ways literacy is embedded in everyday life.

A second concept concerns *context.* Practice-centered approaches challenge definitions of literacy as decontextualized, insisting instead that reading and writing activities are part of a broader range of social activities. Halliday and Martin (1993), for instance, hold that it is necessary to look at the relations between texts and social contexts rather than at texts as decontextualized entities (p. 22). This perspective is useful, for example, in explaining why a book that is a commercial success is not necessarily considered a literary one. Along these same lines, Robinson (1990) notes, "Literacy is impossible to define, for whatever purpose, without reference to its nature and use in some one context—in some one delimited and clearly defined social context" (p. 154). As this passage implies, context is the starting point for understanding reading and writing. Furthermore, looking at context demonstrates the multiple settings that influence reading and writing activities.

A third concept deals with *language norms.* Practice-centered approaches define reading and writing in terms of the normative aspects of language in order to highlight the rules of language use. Lemke (1990), for example, shows that classroom talk about science reflects cultural beliefs about objectivity and subjectivity; Halliday and Martin (1993) make a similar point in their analysis of science textbooks. Likewise, Hymes (1974, 1986) holds that in order to study language, and by extension, reading and writing, it is necessary to examine the rules that are shared by the members of a group, since the participants in a literacy event know how to act as a result of shared patterns of behavior. He also notes that it is necessary to understand the group rules for interpreting reading and writing. This emphasis on rules points out the ways in which people determine what counts as reading and writing within their group and what reading and writing means to group members.

Functionality is a fourth concept. Practice-centered approaches hold that reading and writing is best defined in terms of the functional aspects of language. This definition emphasizes the ways in which people use language to fulfill certain purposes. For instance, in analyzing the language of science textbooks, Halliday and Martin

(1993) suggest that language may serve gatekeeping functions insofar as it restricts access to information. Heath (1983) found that language activities may serve other functions, such as reinforcing community beliefs about child behavior. As a whole, practice-centered approaches draw attention to the actual ways that reading and writing is used in everyday life to achieve specific aims.

Finally, although not universally so, practice-centered approaches recognize the *ideological aspects* of language. The assumption is that language is not neutral. Holding that there is a bias in all definitions of literacy, practice-centered researchers strive to make explicit the value-laden nature of reading and writing. Along these lines, Halliday and Martin (1993) point out the ways in which literacy intersects with class, gender, race, and age. Street (1984, 1993) points to intersections between power and literacy. Taken together, this suggests that even the belief that reading and writing is value-free is in itself a stance.

In sum, practice-centered approaches move literacy research from prescriptive accounts toward descriptive ones, and in doing so, open up many new avenues of investigation. To date, Halliday and Martin (1993) and Lemke (1990) have used systemic functional linguistics, one practice-centered approach, to study science writing. Basso's (1989) study of writing using the ethnography of communication model (Hymes 1974, 1986) suggests yet another method.

HYMESIAN MODEL OF WRITING AS A SOCIAL PRACTICE

The concept of "speech community," defined as a group that shares rules for using and interpreting language, is the basis of the Hymesian model. When Hymes (1974) proposed the concept, his primary unit of analysis was the "speech event." Reading and writing researchers, especially Heath (1983), broadened the definition to include literacy. This new term, "literacy event," refers to any situation in which reading and writing is a central activity. Like the speech event, the literacy event looks at language from eight aspects: *setting, participants, ends, art, key, instrumentality, norms,* and *genre.*

Operationally, setting refers to the time and place of a reading and writing activity, and participants refer to those who are taking part in it. Ends are its expected outcomes or goals. Art refers to form and content; key to manner, tone, or style; and instrumentality to the channel of communication. Norms refers to group rules for using and interpreting language. Genre is the typology of forms. Expanding on this lexicon, Street (1984, 1993) advanced the term "literacy practice." Conceptually, event and practice overlap insofar as a literary practice may be part of several events and a literacy event may include several practices. The concepts *literacy event* and *practice* have proved useful in helping researchers recognize the variety of activities in which people use reading and writing and have helped focus attention on the social settings in which reading and writing are embedded.

A HYMESIAN ANALYSIS OF SCIENCE WRITING AS A SOCIAL PRACTICE

The preceding discussion leads to my main question: Is the Hymesian approach to language applicable to the study and teaching of science writing? By science writing here, I mean any text that deals with science, including scientific reports, the *New York Times* science section, *Discover*, and articles in the *Reader's Digest*. The distinction between scientific writing and science writing is defined as one of function. What is traditionally considered scientific writing is subsumed here under the broader category of science writing.

To explore the research question, I examine four pieces of writing about diabetes. One, by Ronald Kotulak, is from the *Chicago Tribune* and is titled "Thanks to beta cells, doctor can dream about curing diabetes." The second, by Paul Lacy, titled "Pancreatic islet cell transplant," is published in *The Mount Sinai Journal of Medicine*. The third, "Treating diabetes with transplanted cells," also by Lacy, is from *Scientific American*. The last, from the journal of the American Diabetes Association, *Diabetes*, by David W. Sharp and 11 others, with Lacy as the second author, is called "Insulin independence after islet transplantation into type I diabetic patient."

The ethnographer Saville-Troike (1986) notes that not every component of the Hymesian model is salient in every analysis of a literacy event. She also notes that there are varieties of language associated with each component and a series of interrelationships between these components. For example, only particular forms of address are appropriate in specific settings. Moreover, the interrelationships between the components cannot be assumed; they must be discovered. In the case under consideration here, since the examples are texts, the relevant aspects to analyze are participants, ends, art, key, genre, and norms.

PARTICIPANTS, ENDS, ART, KEY, AND GENRE IN SCIENCE WRITING

To analyze science writing with respect to participants means looking at the social relations between the readers and writers of a text. An array of factors may influence these relations. One factor is proximity—that is, the social distance between readers and writers. For example, readers of *Diabetes* and *The Mount Sinai Journal of Medicine* are members of a medical community. In that community, a reader may know a writer, at least by reputation and may judge a writer by his or her affiliation with a laboratory or hospital. In a sense, participants read and write for each other. By comparison, readers of *Scientific American* and the *Chicago Tribune* are educated members of the public. They may also judge a writer by his or her professional affiliation. But the social distance between reader and writer is greater since they are not peers. Additionally, journalists may not hold the same position of authority to the readers as scientists or medical doctors.

Another factor is authorship. The *Diabetes* article lists 12 writers, whereas the other three articles list only one. Readers may wonder about the reasons underlying the particular ordering of names, trying to interpret the list of names according to a hierarchy of relative importance of the authors or some other criteria. This may influence the meaning of the piece for a reader. An additional factor is intertextuality—that is, the relationships between texts. For example, Lacy, who is listed as the second author on the *Diabetes* article, is the primary author of *The Mount Sinai Journal of Medicine* and *Scientific American* articles. Readers interested in the field of transplantation may be familiar with Lacy's works and may draw information from one of his articles to another. His name may influence one's reading.

The point is that the meaning of science writing is constructed by readers and writers. Stated in Hymesian terms, the language writers use in science depends on the readers. From this emerges many questions. Do readers and writers know each other? What is the social position of the readers and writers with respect to each other? What is the position of the writer with respect to his or her community? What is the relationship of the authors in a multiauthored article? What can the reader conclude from a listing of multiple authors? Was writing the only channel for the subject matter? Who and what regulates the writing? As evidenced by this array of questions, the Hymesian framework is useful for highlighting the myriad social relations between readers and writers of science texts.

Regarding ends, science writers may be influenced by a variety of goals, and perhaps some overlapping ones. For instance, in *Diabetes*, Scharp and his colleagues (1990) are reporting the results of their primary research to the readers of the professional society when they note: "This is a preliminary study of the first patient with type I diabetes who has been able to discontinue exogenous insulin after transplantation of islets isolated and purified from cadaver-donor pancreas" (p. 15). However, they are also using the article to acknowledge their funding sources and thank various people who helped them. In different circumstances, these interests may converge or diverge. Lacy's goals in *The Mount Sinai Journal of Medicine* and *Scientific American* are different. The first article (Lacy, 1995), an adaption for publication of his inauguration speech at the Mount Sinai medical school, is a progress report on the state of transplantation research, aimed toward members of the medical community. He starts with the following sentence: "Presented here is an overview of some of the past and present accomplishments in the area of islet transplantation, as well as what I see as developments for the future" (p. 23). The article is his own personal vision of the research. In contrast, the *Scientific American* piece provides general scientific research to the public, as well as a bibliography for further reading. Here he tells the story of the islet transplantation research over the last 25 years. Kotulak's (1995) article in the *Chicago Tribune* points to another purpose: science as news. His article starts as follows:

> His muscles growing increasingly weak and painful after 30 years of diabetes, Steven Craig jumped at the chance to join a medical revolution in treating the disease. Craig,

> who lives in Anaheim, Calif., saw his type I diabetes dramatically reverse after he received a transplant of specially packaged insulin-producing cells that cleared glucose from his body. (p. 1)

The format of the article answers the questions, who, what, where, when, why, and how. By using phrases such as "medical revolution" and "dramatically reverse," the author tells the story of research as a breakthrough in treatment.

Stated theoretically, there are varieties of science writing associated with purpose. Many questions about this connection may be asked. What is the role of science writing among academicians, researchers, consumers, policy makers, and other stakeholders? What is the role of funding in science writing? How is the authority to write about science distributed among the society? How are genre and purpose related? How are genre and topic related? Are there some genres that are inappropriate for some science topics? The scope of these questions suggested by the Hymesian model reveals a range of ways that science writing depends on a writer's goals and emphasizes the centrality of purpose in science writing.

The four articles also differ in the number, type, and placement of illustrations, elements related to the variable of art. The *Diabetes* article contains one table and three multiline graphs. Two of its four pages contain only text. In contrast, the *Scientific American* article has a graphic on each of its seven pages. In addition, the types of graphics differ. The *Scientific American* article contains five colored drawings, one electron micrograph of a human islet, and one photograph of a transplant patient, eating ice cream. By comparison, the *Chicago Tribune* article contains six photographs of various researchers and patients. Graphics in *The Mount Sinai Journal of Medicine* include one electron micrograph of a human islet, four line graphs, one table, and one line drawing of the encapsulation of islets.

The implication is that there are varieties of science writing associated with the aspect of art. This suggests many possibilities. Are some types of graphic communication considered less scientific than others and why? Do some types grant more authority to a piece of writing than others and why? Are some forms of illustration inappropriate in certain settings? Which graphical representations are appropriate with which topics? Clearly, the Hymesian model points out some of the ways in which information about visual and written communication is part of the base of knowledge needed to write and read science.

The four articles are also written in a different key. In *Diabetes*, Scharp et al. (1990) use the passive voice and the third person to describe their work. Transplantation is their subject: "The islet transplantation was performed under local anesthesia with no rise in the portal venous pressure during and after the islet injection" (p. 16). By comparison, Lacy uses first person pronouns in *Scientific American* and *The Mount Sinai Journal of Medicine*. In *Scientific American* (1995), he writes: "Scharp and I decided to place the islets into the network of blood vessels extending from the portal vein in the liver, in part because the system is accessible" (p. 55). The *Chicago Tribune* is more colloquial as it contains quotes from scientists, doctors, and patients. It also uses metaphors of war and sports. Another aspect of key concerns the use of neutral language. The articles that mention potentially divisive topics gloss over controversial aspects of diabetes research. For example, Lacy (1995) writes about the uncertain supply of human pancreases as follows:

> To cope with the shortage, some investigators are exploring implantation of islets from aborted fetuses. They are drawn to this solution partly by the prospect that precursor cells could be isolated and induced to produced islets in quantity. But research into the use of fetal tissues is preliminary. (p. 58)

Missing in this passage is any acknowledgment of the societal attitudes toward abortion in general or fetal tissue research in particular, despite the highly divisive nature of this work. In the article, the problem appears to be a technical one.

As can be seen, an author's use of pronouns, voice, and metaphor suggests that science writing varies with key. This interconnection raises many questions. Which metaphors are appropriate to which types of science writing? Why is the third person considered more objective than the first or second? What purpose does this serve? What purposes do the active and passive voice serve? Here again, the Hymesian model makes explicit the kinds of information one needs to understand science writing.

Finally, the professional science article, the popular science article, and the newspaper account differ in their generic characteristics. Connections between participants, ends, art, key, and genre have already been discussed, but genre blurring emerges as another aspect to consider. Lacy's article in *The Mount Sinai Journal of Medicine*, an adaption of a dedication speech he gave at the medical school, is an example of this. Comparing the speech and the written article would highlight the ways that literacy and orality serve different purposes. However, comparing all the articles shows that there are varieties of science writing associated with genre and raises myriad questions about the generic characteristics of science writing. Which genres are appropriate in which settings? What is the value of each genre? What language variety is suitable for which genre? Addressing these questions helps uncover the tacit knowledge that readers and writers share about science texts. This point about shared knowledge leads directly to a discussion of norms.

NORMS IN SCIENCE WRITING

From a practice-centered viewpoint, part of the job of the researcher is to uncover language norms. In the case of science writing, examples of shared language norms include knowledge of generic characteristics of science texts, as well as knowledge of the uses and functions of science writing. These kinds of knowledge help members of the group make meaning of science writing.

With respect to generic characteristics, readers and writers must share a typology and taxonomy of science genres. Knowing the typology means that readers and writers can differentiate between a wide range of science genres. They know how to distinguish scientific reports, proposals, science papers, journal articles, review articles, popular science articles, news magazine articles, popular magazine articles, environmental impact statements, public service announcements, brochures, and other material. Readers and writers also have to share a taxonomy of these genres. They have to know that each genre serves a different function and has a different value within their speech community. For example, readers know that the account given in the *Chicago Tribune* does not have equal status as the description of the same technique found in *The Mount Sinai Journal of Medicine* article.

Regarding the uses and functions of science writing, readers and writers have to know what counts in different situations. For example, even though prescriptive guidelines for science writing state that narrative language is not appropriate for science writing, readers know that Lacy's (1995) *Scientific American* article is a story about the history of transplantation research in the United States. A reader deduces this from the author's first words: "Until about 75 years ago, the form of diabetes that usually strikes children and young adults was invariably fatal" (p. 50). From this single sentence, the reader recognizes that a story will follow. The writer goes on to relate the story of the field, making it dramatic and suspenseful. In order to accept this as a science text, readers and writers have to share the group's rules for writing about science. In addition, readers and writers have to share the knowledge that science writing may serve a number of purposes. This knowledge must be part of one's repertoire when reading and writing about science. And finally, readers and writers have to share rules for interpreting science writing. Taken together, such knowledge is the basis of the group's language norms for science writing.

In sum, without an awareness of norms, researchers cannot understand the meaning of science writing—and by extension, of literacy events or practices—to a speech community. Focusing on norms has implications for research and pedagogy.

IMPLICATIONS FOR RESEARCH

Just as genre theorists catalog literary and folk genres, literacy researchers need to catalog science writing and genres. Cataloging this material will be a step toward developing a typology and taxonomy of science writing and will help researchers analyze science writing as a set of literacy events and practices.

In addition to looking at science writing as a textual practice, researchers need to look at science writing as a set of social practices. Schools, research centers, funding agencies, laboratories, publishing houses, and legislatures all need to be analyzed. Findings from these studies would help to reveal norms of interpretation. These studies would help researchers identify the repertoire of reading and writing strategies needed to interpret science writing. Ultimately, the communicative competence needed for science writing could be outlined.

Additionally, science writing has a historical aspect. Although science writing practices and genres may be analyzed separately, they may also be seen as part of an overall literacy tradition, suggesting that science writing is part of a community's literacy practices and associated genres. These in turn are part of a larger knowledge base. Gee (1990) postulates that knowledge about literacy may not be equally shared by all members of a community, but that every individual has access to the community's system of making meaning through language. Thus, science writing needs to be examined from this macro level.

IMPLICATIONS FOR PEDAGOGY

Implications exist for the teaching of science writing. First of all, we should present science writing as a set of multiple practices, making distinctions between language varieties and science genres. To do this, we can use models of science writing that highlight varieties and genres. For example, in my technical writing class, I ask students to find an article in one journal to rewrite for another one. As the students rewrite the article, they are confronted by matters of organization, detail, voice, graphics, and a host of other writing issues. In grappling with these matters, they discover what counts as science writing in various settings. They see that science writing does not constitute a single set of uses, functions, and meanings. In essence, this is what they know from their experiences as readers and writers.

Secondly, if we are to use practice-centered approaches in our classrooms, we need course materials that incorporate this new perspective. Popular texts, such as Borowick (1996), Markel (1996), Pattow and Wresch (1993), and VanAlstyne and Maddison (1994) tend to gloss over the functional aspects of science writing. One notable exception is Weisman (1992). To address this situation, we could design new materials and make them available in print or online. This would be a step toward adopting methods of teaching science writing that are informed by the view that science writing, like writing in general, is a set of social and cultural practices.

REFERENCES

Basso, K. (1989). The ethnography of writing. In R. Bauman & J. Sherzer (Eds.), *Explorations in the ethnography of speaking* (2nd ed., pp. 425–432). Cambridge: Cambridge University Press.

Borowick, J. N. (1996). *Technical communication and its applications*. Englewood Cliffs, NJ: Prentice Hall.

Gee, J. (1990). *Social linguistics and literacy: Ideology in discourses*. New York: Falmer.

Halliday, M. A. K., & Martin, J. R. (1993). *Writing science: Literacy and discursive power*. Pittsburgh, PA: University of Pittsburgh Press.

Heath, S. B. (1983). *Ways with words*. Cambridge: Cambridge University Press.

Hymes, D. (1974). *Foundations in sociolinguistics: An ethnographic approach*. Philadelphia: University of Pennsylvania Press.

Hymes, D. (1986). Models for the interaction of language and social life. In J. J. Gumperz & D. Hymes (Eds.), *Directions in sociolinguistics: The ethnography of communication* (pp. 35–71). Oxford: Basil Blackwell.

Kotulak, R. (1995, February 7). Thanks to beta cells, doctor can dream about curing diabetes. *Chicago Tribune*, p. 1.

Lacy, P. E. (1994). Pancreatic islet cell transplant. *The Mount Sinai Journal of Medicine, 61*(1), 23–31.

Lacy, P. E. (1995). Treating diabetes with transplanted cells. *Scientific American, 273*(1), 50–58.

Lemke, J. L. (1990). *Talking science: Language, learning, and values*. Norwood, NJ: Ablex.

Markel, M. (1996). *Technical communication: Situations and strategies* (4th ed.). New York: St. Martin's Press.

Pattow, D., & Wresch, W. (1993). *Communicating technical information: A guide for the electronic age*. Englewood Cliffs, NJ: Prentice Hall.

Robinson, J. (1990). *Conversations on the written word: Essays on language and literacy*. Portsmouth, NH: Boynton/Cook.

Saville-Troike, M. (1986). *The ethnography of communication: An introduction*. Oxford: Basil Blackwell.

Scharp, D. W., Lacy, P. E., Santiago, J. V., McCullough, C. S., Weide, L. G., Falqui, L., Marchetti, P., Gingerich, R. L., Jaffe, A. S., Cryer, P. E., Anderson, C. B., & Flye, M. W. (1990). Insulin independence after islet transplantation into type I diabetic patient. *Diabetes, 39*(5), 15–18.

Street, B. (1984). Literacy in theory and in practice. Cambridge: Cambridge University Press.

Street, B. (1993). *Cross-cultural approaches to the study of literacy*. Cambridge: Cambridge University Press.

VanAlstyne, J. S., & Maddison, G. R. (1994). *Professional & technical writing strategies: Communicating in technology and science* (3rd ed.). Englewood Cliffs, NJ: Prentice Hall.

Weisman, H. M. (1992). *Basic technical writing* (6th ed.). New York: Macmillan.

About the Contributors

Dwight Atkinson

Dwight Atkinson is an applied linguist in the English Department at Auburn University. His research deals with interrelationships between writing/literacy and culture, treating such topics as the social history of medical and scientific research writing, differences between the "cultures" of first and second-language writing instruction, critical thinking as a sociocultural construct, and using (both separately and in combination) qualitative and quantitative research methods to understand literate practices in context. A book-length study on the sociohistorical development of discourse in the *Philosophical Transactions of the Royal Society of London* is scheduled for publication by Erlbaum in 1998.

Philippa Jane Benson

Philippa Jane Benson is the Coordinator of Writing Instruction in the Department of Biological Sciences at the University of Pittsburgh. She is also the editor of the *Bulletin of the Association of Teachers of Technical Writing*. Dr. Benson has been teaching technical and scientific communication in academia and in industry for over fifteen years.

John H. Carpenter

John Carpenter is a Ph.D. candidate in Rhetoric and Technical Communication at Michigan Technological University, where he teaches technical communication, composition, and speech communication. His research lies in rhetorical theory and criticism, with an emphasis on the role of scientific rhetoric in environmental controversies. Carpenter is currently undertaking dissertation research that reexamines the debate between resource conservation and wilderness preservation, focusing on the scientific writings of John Muir and Gifford Pinchot.

Steven Darian

Steven Darian is Professor of Communication and Education at Rutgers University, Camden, New Jersey, where he is Chairman of the Education Department and director of the certification program in Teaching English as a Second Language. His specializations include English for Science and English for Business. He has taught ESL and English for Specific Purposes in Afghanistan, Saudi Arabia, Hungary, and India—where he served as a Fulbright Senior Lecturer in 1992. His articles have appeared in *The Modern Language Journal*, *ESP Malaysia*, *TESOL Quarterly*, *IRAL*, *Fachsprache*, *System*, and *The Journal of Pragmatics*. He is spending the 1997-98 academic year as a Fulbright scholar in Tashkent and Samarkand.

Dan Ding

Dan Ding is a doctoral candidate in the Technical Writing Program of the Department of English, Illinois State University. He is teaching English 101 and an academic writing course. He is also an evaluator of the University Writing Examination Program. The idea of this chapter came to him when he was reviewing some 19th-century issues of the *Philosophical Transactions of the Royal Society of London*, in an attempt to choose some articles for his academic writing course. Currently, he is working on his dissertation, which is about the relationship between scientific discourse as it exists in college and university classrooms and scientific discourse as it exists in professional journals.

Gay M. Gragson

Gay M. Gragson is a technical writer in the Department of Genetics at the University of Georgia and coordinating editor for the Editor-in-Chief of *GENETICA*. She holds a master's degree in technical writing from The Pennsylvania State University. For over a decade she has also assisted the Executive Director of the Office of Computer and Information Systems at Penn State in creating strategic planning and technical documents.

Ted L. Gragson

Ted L. Gragson is an assistant professor in the Department of Anthropology and faculty member in the Conservation and Sustainable Development Program of the Institute of Ecology at the University of Georgia. He holds a doctoral degree in anthropology from The Pennsylvania State University. He has disciplinary and interdisciplinary teaching obligations in which he enjoys introducing students to the theory and method of ecology. He currently has research projects in Paraguay, the Dominican Republic, and Southern Appalachia focusing on human dimensions of natural resource use and environmental transformation.

Cynthia R. Haller

Cynthia R. Haller is an Assistant Professor of English at North Carolina State University, where she teaches business and technical communication and serves on the faculty of the MS in Technical Communication program. She is currently examining how scientific and technological information has historically been communicated to farmers and farm women through agricultural extension services, and is also interested in connections between rhetoric and engineering design.

Rosemary Horowitz

Rosemary Horowitz is an assistant professor of English at Appalachian State University, a unit of the University of North Carolina. She is interested in the study of literacy overall. Specific research areas include writing and the teaching of writing, theory and practice of translation, and writing in the Jewish community. Her book on the writing of Jewish survivors of the Holocaust, *Literacy and cultural transmission in the reading, writing, and rewriting of Jewish memorial books*, is scheduled to be published by Austin & Winfield.

Richard D. Johnson-Sheehan

Richard D. Johnson-Sheehan is a Professor of Language and Rhetoric at the University of New Mexico. He has published numerous works on metaphor in the rhetoric of science. He primarily studies the use of metaphor and narrative in the invention and development of quantum mechanics, relativity theory, and complexity theory.

Carmen Pèrez-Llantada Auria

Carmen Pèrez-Llantada Auria works as an associate professor of ESP at the University of Zaragoza, Spain, where she has been teaching since 1990. In addition to her articles on interdisciplinary studies between 20th century science and contemporary literature, she is currently engaged in pragmalinguistic research and has published several essays concerning discourse theory, sociolinguistics and methodology in the field of EST.

Trevor Pinch

Trevor Pinch is Professor in the Department of Science and Technology Studies, Cornell University. He has published widely on the sociology of science and the sociology of technology. His latest book, *The Hard Sell*, London: HarperCollins, 1996 (co-authored with Colin Clark) is on the rhetorical skills of market sellers and street hustlers. He and Harry Collins are currently completing for Cambridge

University Press a second volume of *The Golem* series popularizing the field of Science and Technology Studies.

Ramòn Plo Alastruè

Ramòn Plo Alastruè works as an associate professor of English for Science and Technology at the University of Zaragoza, Spain. His research interests include science writing, intertextuality and pragmatics. Apart from his work in the area of EST, he is currently working on the subject of structure and ideology in contemporary narrative.

Mary Rosner

Mary Rosner is associate professor of English at the University of Louisville. She teaches undergraduate and graduate courses in rhetoric and in Victorian fiction, and she is currently studying how discourses from law and psychology compete for authority in the Victorian courtroom.

Christine M. Skolnik

Christine M. Skolnik is a senior Ph.D. candidate in Rhetoric and English Studies at The Pennsylvania State University. She is currently working on her dissertation, *Rhetorics of the Sublime*, which examines intersections between eighteenth-century empiricisms and aesthetics. Ms. Skolnik also teaches technical writing and conducts research into the rhetoric of contemporary scientific discourse.

Charles Alan Taylor

Charles Alan Taylor (Ph.D., University of Illinois, 1990) is Associate Professor of Communications at Drury College, Springfield, Missouri, where he teaches courses in persuasion and rhetorical theory and criticism. His research concerns critical analysis of scientific rhetoric, rhetorical epistemology, and the interface of technical reasoning and public deliberation.

Author Index

Subject Index